¡Saludos, castores!

1

MARAVILLAS ANIMALES 17

LOS CASTORES

QUINN M. ARNOLD

CREATIVE EDUCATION | CREATIVE PAPERBACKS

¿QUIÉN ES UN CASTOR OCUPADO? YO.

Índice

¡Saludos, castores! 1
Constructores ocupados 7
La cola del castor 8
Dientes para los árboles 10
¡Hora de comer! 12
Las crías del castor 14
¿Qué hacen los castores? 16
¡Adiós, castores! 18
Imagina un castor 20
Palabras que debes conocer 22
Índice 24

Publicado por Creative Education y Creative Paperbacks
P.O. Box 227, Mankato, Minnesota 56002
Creative Education y Creative Paperbacks
son sellos editoriales de The Creative Company
www.thecreativecompany.us

Diseño de Graham Morgan
Dirección artística de Blue Design (www.bluedes.com)

Imágenes de Alamy/Imagebroker, 14-15; Dreamstime/Chase Dekker, 6-7, Jnjhuz, 12-13; Flickr/Biodiversity Heritage Library, 2; Getty Images/Paul A. Souders, 4; Minden Pictures/Dietmar Nill, 1, Konrad Wothe, 8-9; Newscom/imageBROKER/Bernd Zoller, 17; Pexels/Denitsa Kireva, portada (derecha); Public Domain, 16, 23; Public Domain/Biodiversity Heritage Library, portada (centro); Shutterstock/Capitan Crizelini, 18-19, Christian Musat, 24, S_Images.com, 3, 20-21, Thyrymn2, portada (izquierda); Wikimedia Commons/Sylvain Haye, 11

Library of Congress Cataloging-in-Publication Data
Título: Los castores / por Quinn M. Arnold.
Names: Arnold, Quinn M., author.
Title: Los castores / by Quinn M. Arnold.
Other titles: Beavers. Spanish
Description: Mankato, Minnesota : Creative Education and Creative Paperbacks, [2025] | Series: Maravillas | Includes index. | Audience: Ages 4-7 | Audience: Grades K-1 | Summary: "An introduction to beavers, this beginning reader features eye-catching photographs, humorous captions, and basic life science facts about these wetland animals. This Spanish text includes a labeled image guide, glossary, and index"-- Provided by publisher.
Identifiers: LCCN 2024021896 (print) | LCCN 2024021897 (ebook) | ISBN 9798889895015 (library binding) | ISBN 9781682777121 (paperback) | ISBN 9798889895138 (ebook)
Subjects: LCSH: Beavers--Juvenile literature.
Classification: LCC QL737.R632 A7618 2025 (print) | LCC QL737.R632 (ebook) | DDC 599.37--dc23/eng/20240627

Impreso en China

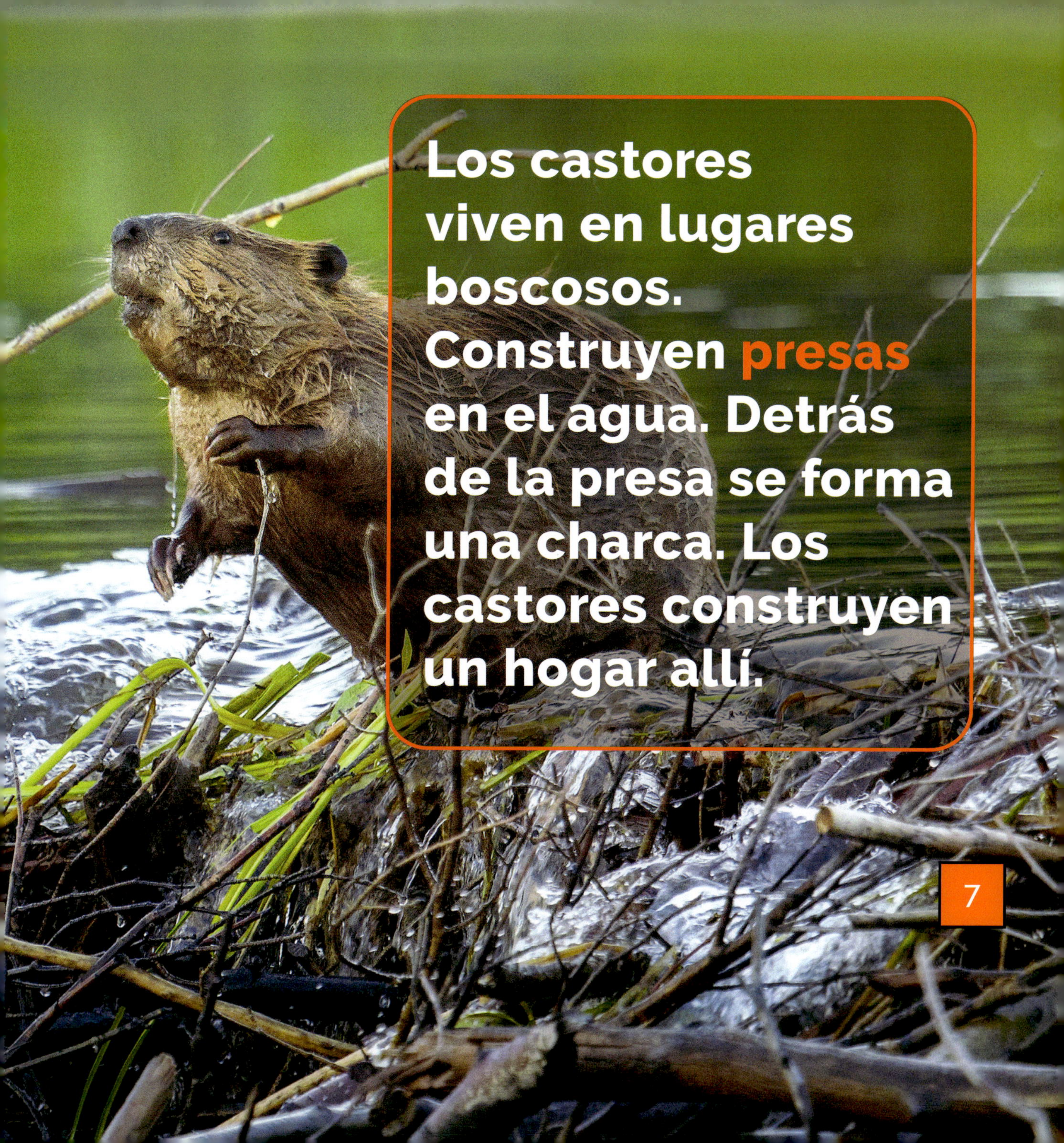

Los castores viven en lugares boscosos. Construyen **presas** en el agua. Detrás de la presa se forma una charca. Los castores construyen un hogar allí.

MI PISCINA ES MÁS GRANDE QUE LA SUYA.

El pelaje del castor puede ser café o negro. La cola es grande y plana. Las patas traseras de un castor son **palmeadas**. Esto le ayuda a nadar.

Los castores tienen grandes dientes delanteros. Son afilados y de color naranja.

Los castores mastican los árboles. Utilizan los árboles de la presa.

LOS DIENTES DEL CASTOR SIEMPRE ESTÁN CRECIENDO.

Los castores hambrientos comen madera y la corteza del árbol. También comen plantas. Los castores almacenan comida para el invierno.

¿QUIERES UN BOCA-DO?

LAS CRÍAS DEL CASTOR PUEDEN NADAR A LOS CUATRO DÍAS DE VIDA.

Las crías del castor se llaman "kits". Beben leche de su madre. Las crías permanecen con su familia unos dos años.

Los castores ocupados trabajan en sus presas. Recogen comida por la noche. Se sumergen bajo el agua.

NADA
COMO UN BAÑO
DESPUÉS DE UN DÍA DE
TRABAJO MUY DURO.

¡Adiós, castores!

[Imagina un castor]

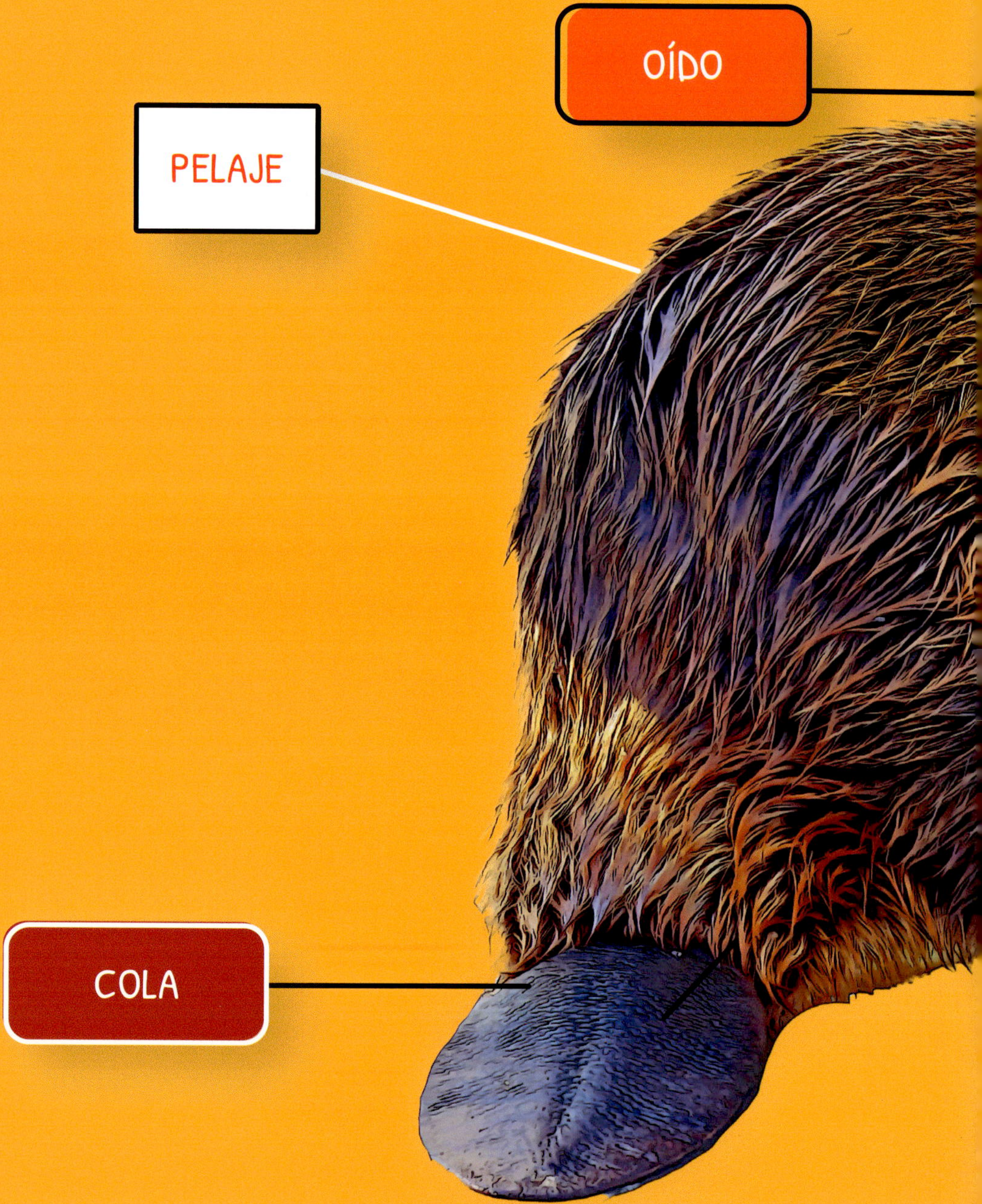

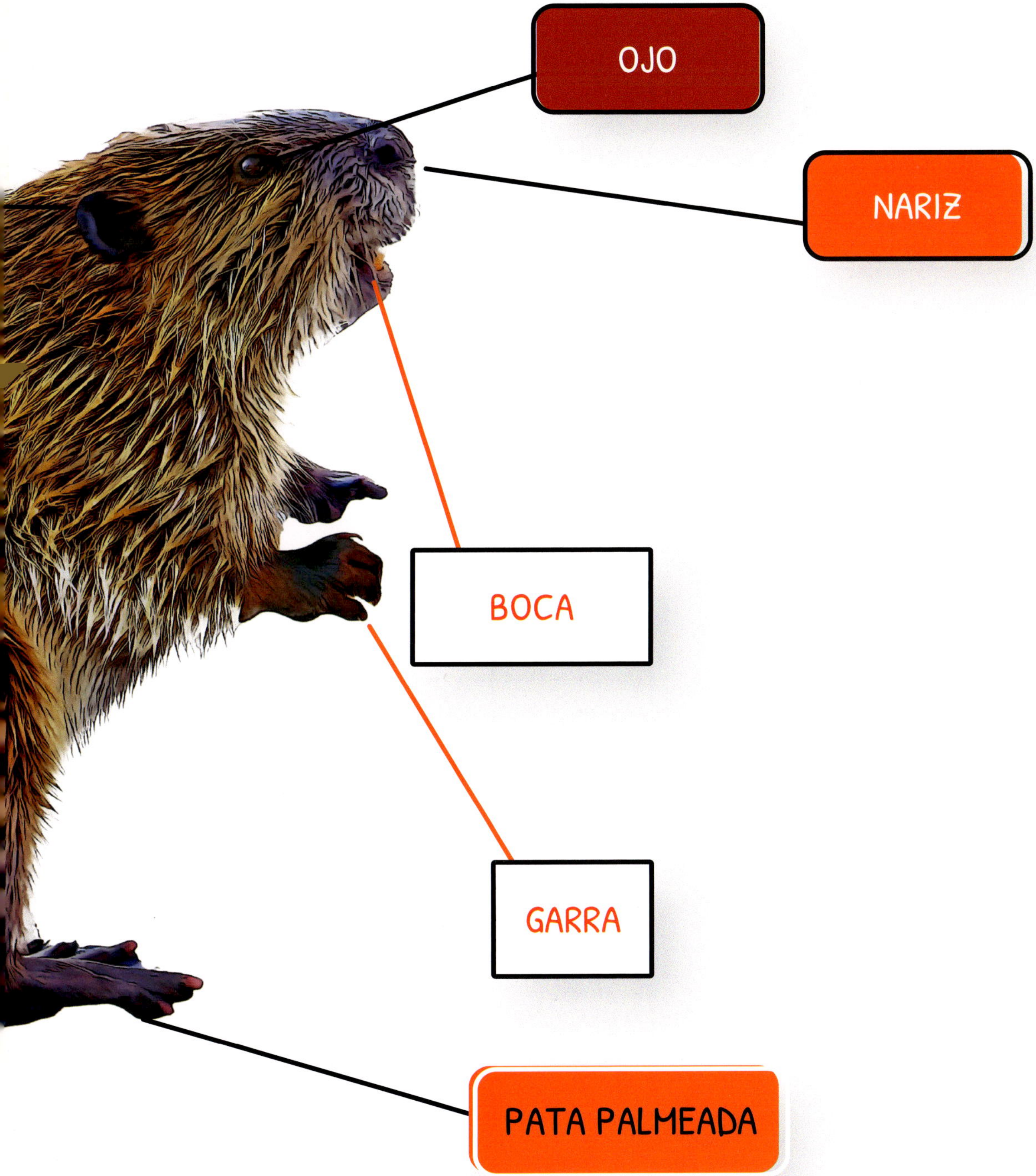
OJO
NARIZ
BOCA
GARRA
PATA PALMEADA

PALABRAS QUE DEBES CONOCER

corteza: la dura cubierta exterior de una planta leñosa

palmeada: conectada con la piel

presa: una barrera a través de un río o arroyo hecha de árboles, ramas y otros materiales naturales

BEAVER

ÍNDICE

alimentación, 12, 15, 16

colas, 9, 20

charcas, 7

crías, 14, 15

dientes, 10, 11

natación, 9, 16

patas, 9, 21

pelaje, 9, 20

presas, 7, 10, 16

pson, Boris Williams and Perry

niversal Music – MGB Songs

547-8979.

ins Children's LLC. Harvest
emark.

.PC

LISHERS

5 5 4 3 2 1

Cover design by Faceout Studio, Spencer Fuller
Cover images © Rawpixel.com / Shutterstock
Interior design by Wildwood Digital Publishing

For bulk, special sales, or ministry purchases, please call 1-800-…
Email: CustomerService@hhpbooks.com

This logo is a federally registered trademark of the Hawk… House Publishers, Inc., is the exclusive licensee of this trad…

In the Valley of the Shadow
Copyright © 2026 by Amy Lesslie Kulbock
Published by Harvest House Publishers
Eugene, Oregon 97408
www.harvesthousepublishers.com

ISBN 978-0-7369-9247-3 (Hardcover)
ISBN 978-0-7369-9248-0 (eBook)

Library of Congress Control Number: 2025937067

Printed in Colombia

25 26 27 28 29 30 31 32 33 34 / NI / 10 9 8 7 …

To my mom, Barbara Lesslie,
who showed me what it means to use
your life to point to Christ.

To my dad, Robert Lesslie,
who turned dreams into reality and taught
me that I could do the same.

To my niece, Adah Lesslie,
whose love for exploring and understanding God's creation
reminds me to engage the world with wonder.

To my nephew, Noah Lesslie,
who used everything he had to make people smile.

A Psalm of David

The LORD is my shepherd; I shall not want.
He makes me lie down in green pastures.
He leads me beside still waters.
He restores my soul.
He leads me in paths of righteousness
for his name's sake.
Even though I walk through the valley of
the shadow of death,
I will fear no evil,
for you are with me;
your rod and your staff,
they comfort me.
You prepare a table before me
in the presence of my enemies;
you anoint my head with oil;
my cup overflows.
Surely goodness and mercy shall follow me
all the days of my life,
and I shall dwell in the house
of the LORD forever.

Psalm 23 ESV

Contents

Introduction 9

1. On Suffering and Grief:
Navigating the Inevitable 13

2. Reeling from Impact:
Understanding Shock and What it Takes
to Get Both Feet Back on the Ground 27

3. The Vessel: Why Your
Worldview Is So Important 39

4. Help: Why It Is Vital to
Invest in Your Relationships Now 49

5. Making Room for God:
How to Get out of the Way 63

6. Tackling the Unthinkable:
Traversing the Waters of Forgiveness 75

7. Regret: Facing Hard Things,
Owning Them, and Letting Go 87

8. Loving Your Neighbor: A Closer Look
at Our Jobs as Caregivers and Grievers. 99

9. What Now? The Discipline of Simplicity 113

Conclusion: Looking to the Future 127
Voices from the Couch on Suffering and Grief. 137
My Reflections 147
Notes 155

Introduction

We never know what tomorrow holds, but we know the One who holds tomorrow.

Notes from the Bible of Barbara Lesslie

My life took a tragic and sudden turn several years ago. The shaping of my life post-tragedy is ongoing, and I expect the growth process to continue as long as I breathe. The concept of this book, though, existed in some form before that tragic day. As a mental health clinician, I journey with those who are trying to make sense of and do something with their suffering. I get to climb down into the sacred trenches of pain, and I have learned a great deal from that trench work.

Over the years it's become clearer to me that grief is a human conflict. And as a conflict, it is frustratingly unresolvable. Even if we are facing it head-on, we won't be able to negotiate it out of our stories or erase the loss. Instead, grief is something we must manage. But what we are managing at any given moment can evolve. Grief does not go away, but it changes form.

If grief has stages, I have yet to make sense of them. The word *stages* indicates discernible movement from one thing to the next. A progression. But I can't find a line between sadness and anger, or shock and acceptance. I can't find a clear path, and grief certainly doesn't come with a map. When wrestling with grief, it seems like everything and nothing makes sense at the same time. There are places in grief where the most precious human experiences emerge like bright light. There are also dark places where even our own skin feels unfamiliar. Maybe it is this dynamic we need to explore: the relationship between the light and the shadow in grief. They are both real and woven together into the fabric of our earthly wrestling with loss.

My experience as a licensed professional counselor has taught me that there are no rules in grief. There is, unfortunately, no predictable timeline or roadmap for getting to the other side either. There is also no definitive picture of what the other side should look like. We do know that change is inevitable, however. More specifically, change is unavoidable. Change can happen instantly and come without invitation or consent. Grief, in part, is surviving the initial change that occurs with the impact of loss, wading through the ripples of damage, and, finally, settling into a new normal. This process has been reinforced in my personal life as well as my professional life. Now I travel alongside my family and friends as we grieve the tragic loss of my parents, niece, and nephew to a mass shooting. Our personal grief experience has also reinforced the idea that one of the most valuable parts of walking through grief is not getting to the other side, but honoring the value of every step. And I do mean every single step.

This is where the light and the shadow converge. They wrestle and dance in the now of every step we take in the journey of grief. If we submit to this awareness, our job becomes less about making sense of things and more about exposing ourselves to the surprises of this amazingly divine experience. The Christian experience of grief is particularly focused on hope, acceptance, and meaning. Where the light and shadow converge, there God is also. He is in it, holding us up, molding us into His image.

Words are a powerful catalyst. As a Christian I believe there is a treasured holiness to words that can breathe light into darkness. Part of what I do professionally is help others develop the language that shapes their experiences because their words are that important. Creation begins with the words of God, and we are sustained through the gift of His Word to us in Scripture. We are also tasked with guarding and honoring our own words. We are made in His image, and our words are powerful indeed.

Many words have previously been written about suffering and grief. It is not my intention to reinvent the wheel, but to impart what I have learned through the words of others, and through the shaping of my own story. My hope is to help you explore and define your own experience in suffering. There is a section in the back of the book called "My Reflections" where you can write down your thoughts. If my words encourage you to seek God, find meaning, and reveal the depth of your experience amid life and loss, I am honored to play a part in that journey.

CHAPTER 1

On Suffering and Grief

Navigating the Inevitable

Will the Lord spurn forever, and never again be favorable? Has his steadfast love forever ceased? Are his promises at an end for all time? Has God forgotten to be gracious? Has he in anger shut up his compassion?

Psalm 77:7-9 ESV

In random order of importance, I am a licensed professional Christian counselor, a mom, a sister, a business partner, a friend, a supporter of the arts, an adventurer, a Bible teacher for adults with special needs, a volunteer, an amateur bird enthusiast, a mountain lover, a wife, and a daughter. I am also someone lucky enough to have known who I am and who I want to be for as long as I can remember. Periods of struggle notwithstanding, I have faced life with the assistance of a solid sense of identity. God knit me together with His sovereign hands this way. He knew what was coming.

In April 2021, the property where my home rests was forever marred by something unthinkable. My neighbors, otherwise known as my parents, were killed along with my niece and nephew, and two HVAC workers who were working at the house. It was a beautiful spring day. It was a quiet, sunny, unremarkable afternoon. It was a shattering. A mass shooting by a stranger. It happened in the sanctity and safety of their home. It was an undoing.

At the time of this writing, nearly four years later, those words still seem wrong. It is still surreal and jarring to acknowledge. Living through the shattering impact of trauma and then the long, slow voyage of grief has been the hardest thing I've ever done, and it has also gifted the most sacred blessings. There are inherent paradoxes in Christian grief. There is loss and gain. Anger and peace. Isolation and connection. Confusion and purpose. Anguish and grace. Reconciling these is part of the work we do when called to bear this particular cup. Weaving aspects of my personal story with my professional experience and my Chistian faith through this book is one of the ways I am reconciling these paradoxes for myself. I recognize the chaos that is suffering and grief, and I am determined to try to forge a path of meaning through it.

Grief Accessibility

In the spring of 2024, when I was beginning preparations to speak at a Christian conference for young adults, a teenage girl asked me to speak on the topic of suffering. She made clear that she thinks her generation is hungry for not only recognition of their unique set of struggles, but tools to help them cope. Youthful resilience is not adequate for the task. Their age does not protect them from the condition of suffering that requires very grown-up responses. The fact that our world is in such a state that our children would ask to be educated on dealing with suffering is something to grieve on its own. She was right, though. Not only is our world dishing out suffering by the heap, but we are broadcasting every ounce of it to one another in real time. We not only have to hold our own experiences of suffering and grief, but we are also inundated with the global, regional, and individual sufferings of our neighbors. It is heavy.

This kind of grief accessibility is, like so much of our culture, complicated. There are benefits and drawbacks to being able to access the real-time stories of people all over the world. On the plus side, the intricate context that history books miss is being made complete before our eyes. We not only see world issues unveil, but we know how people from all sides are feeling and experiencing them. The telling of current events is no longer in exclusive hands, but in the hands of everyone via the internet and social media. We are getting a more complete picture of history in the making, of suffering, than ever before.

Often, because we hear of and read about them, we can respond to these stories. We can tend to what we can see. And now we can see just about everything. There is good here. The opportunities for intervention and ministry seem to be infinite and accessible. Because we have access to understanding the nuance of specific needs, beyond broad strokes, we can respond with more precision. From the comfort of our phones or desks, we can organize money, goods, and services across the world. Service opportunities have never been so advertised and diversified. This is also true on an interpersonal level. We might read about a tsunami on the other side of the world in the same minute our neighbor posts online about her catastrophic medical diagnosis. While there is certainly ample opportunity via the internet for deception and manipulation, our ability to respond to real need is acute and available.

The access we have to stories of suffering can also have a positive impact on our perspective. Grief and suffering have an isolating effect. These conditions tend to drive us into our own heads and homes, where it is easy to be convinced that we are alone and that our experience can't possibly be understood. We are in an information age that has transformed our ability to connect with those who share our life experiences no matter where they are. It is often so helpful to read about or talk to someone whose story mirrors our own. We can also gain perspective by being exposed to stories that challenge our perceptions. These might be stories of hope when we are hopeless or help when we feel helpless. They might be stories of suffering that remind us of our blessings. The power of perspective cannot be understated, and making the world smaller through communication can impact the way we view our stories. Our ability to know about and respond to the needs of our communities is a blessing and a privilege.

It can also be overwhelming. As a mental health counselor, I deal with the impact of suffering and grief daily. A client showed up recently with an acute panic attack triggered by her inability to tend to the suffering in the Middle East. She was overloaded with guilt about not being able to respond to the hurt being broadcast across the news and social media. She felt unable to escape the constant thrum of need around her and across the world. I get it. The weight of community and global need is heavy, and responding to it gets trickier when it is competing for space with our own suffering and grief. On the flip side, the constant inundation of suffering can simply make us numb to the needs around us. The truth is, we simply can't personally attend to all the needs of the world or even those of our own communities or families. It is a humbling recognition that can lead to compassion fatigue even in the most fortitudinous of us.

The Looming Question

But here we are, engaging together in a journey about grief and suffering because we are struggling with pain. We are trying to figure out how best to tend to our pain and the pain of those around us. Right out of the gate, we can acknowledge that grief and suffering are complicated topics that bring up a lot of logistic and existential questions. I knew that when I spoke at the conference for Christian teenagers, I would have to address the one big question that tends to come up whenever we talk about suffering and grief: Why? Why is the world like this? Why did this happen? If God is good, why is there pain? If God loves me, why won't He take this cup from me? Why is there suffering if God is omnipotent, omniscient, and omnibenevolent? I cannot sufficiently answer this. What I am attempting to do, however, is break it down into ideas that can help us navigate this giant question: WHY?

W: Who am I amid suffering?
H: How can I move through the world carrying this thing I now know?
Y: Yield to the movement of God.

Who Am I Amid Suffering?

First, who am I amid this pain? Suffering is largely defined by our experience of it. Who we are shapes the way we interpret our experience. Pain does not exist in a vacuum. In order to be felt, there has to be a collision; the pain has to land on something. And, unfortunately, you are the point of impact. When crisis happens, the results can range from discomfort to disfigurement (literally and figuratively). While the components of a crisis are typically out of our hands, we do have some control over the state we are in when it hits.

Suffering is largely defined by our experience of it.

Some of the pieces that come together to build the foundation of self are culture, genetics, life experience, and environment. But I believe that the cornerstone upon which *self* sits is our worldview. Unlike so many factors that are out of our control, we get to participate in and make choices about our worldview. It is an exercise in the understanding of identity. We have some power here. If we so choose, we get to actively engage in the process of determining what we believe. In chapter 3 of this book, I flesh out the idea of worldview in greater detail. For now, it is enough to conceptualize that the fullness of suffering and our reaction to it is in large part determined by the shape of its receiver.

The nature of living in a sinful world ensures that pain will happen in and around us. We are not promised a life free from suffering. No amount of good works will elevate us above the fray of fallenness. We cannot control the way it sometimes pours into our lives, wreaking destruction like a flood. While I am not suggesting that who you are, or your identity, determines how great or small your suffering is, I am suggesting that the foundation of self gives shape to your pain. It is the source from which you ascertain what has meaning, what is good or bad, what is valuable or worthless, and what is worth holding on to. These are fundamental tethers amid life's storms.

No amount of good works will elevate us above the fray of fallenness.

How Can I Move Through the World Carrying This Thing I Now Know?

When the dust of our trauma begins to settle, and the clarity of the destruction declares itself, it is up to us to figure out what to do with the new landscape of our lives. How can I even begin to consider getting back to normal? What am I supposed to do when normal is no longer an option? Am I even capable of living in a way that will bring joy or fulfillment? How can I go out into a community that doesn't seem to fit me anymore? Who am I after all this? How do I move through the world knowing what I now know about it?

Grief changes us. Studies show that grief not only impacts our mood and perceptions of the world, but can actually change our brains and physiology.[1] It can impact our cardiovascular systems every bit as much as it alters our sleep patterns and eating habits. We might be able to easily define or put simple words to our initial loss, but the ripples of grief extend far beyond the point of impact. It can be hard to recognize ourselves when the dust settles.

I often think of my life before grief as being characterized by certain privilege. Even though I cognitively knew better, there were people and experiences I took for granted. There were inconsequential things I elevated to important and important things I minimized so much they became unseeable. When challenged, I tried to pay attention to these maladaptive tendencies. But the comforts and busyness of life often got in the way of self-motivated challenge. We typically do not consider what we haven't been confronted by yet. We don't exactly ask for this kind of change. If you have experienced suffering, trauma, or grief, you understand the nonconsensual ripping away of comfortable ideas and ways of living. We are forced into knowing things we did not choose to know. In so many areas, we lose the privilege of comfortable ignorance.

To be honest, the human in me really doesn't want to know the things I now know about the world because of pain. Through my job and my own experience, my eyes are opened to a kind of suffering that I have a hard time making sense of. I don't see it all (thank you, Jesus) but I see enough to recognize the dark, heavy pull of the knowledge of what man is capable of. I am aware of an unthinkable depth of depravity, and also aware of our immeasurable capacity to absorb the pain it causes.

How do we move through the world knowing these dark and confusing things about it? Suffering with purpose requires a confrontation with this question. Surviving that confrontation is, in one sense, the journey of grief itself. Several chapters in this book tackle aspects of this exploration. What is important to note, though, is that asking ourselves this question is a *choice*. This is work we do not have to do. We can, instead, allow our pain to craft an entirely new set of blinders for us to hold on to. Confronting the *whats*, *hows*, and *whys* of grief and suffering is choosing to open our eyes wide to the truths of *who* and *where* we are.

Yield to the Movement of God

This brings us to the Y of WHY. What does it look like to yield to the movement of God? This question bears some consultation with theology. Rooted at the center is the discipline of trusting that God is who He says He is. This is not as easy as it reads (maybe discipline never is). This involves not just believing in the identity of Christ as the only Son of God and the source of our redemption but believing and behaving in a way that confirms that He is all the other things as well. Here is a list of some of the characteristics of God found in Scripture:

- God is a rock (Psalm 18:2).
- God loves you (Romans 8:38-39).
- God is kind (Psalm 145:8-9).
- God is all-powerful (Isaiah 44:6–48:11).
- God pays attention to even you (1 Peter 5:7).
- God is active (Hebrews 4:12).
- God cares about justice (Psalm 11:7).

- God delights in your well-being and happiness (Psalm 35:27).
- God is faithful (Deuteronomy 7:9).
- God is sufficient to meet your needs (Philippians 4:19).
- God is a shield (Psalm 3:3-6).
- God is sovereign (Matthew 10:29-31).
- God is creative (Genesis 1:1-31).
- God is everywhere, all the time (Jeremiah 23:23-34).
- God is mysterious (Ecclesiastes 11:5).
- God is merciful (Lamentations 3:22).
- God is eternal (Romans 1:20).
- God has a plan (Ephesians 1:11).
- God's plan is perfect (Psalm 18:30).
- God is always good (1 John 1:5).
- These aspects of God never change (Malachi 3:6).

What does it mean if I actively choose to believe that all these truths about God are accurate? For starters, it means that I am relinquishing control. I am humbly submitting to the idea that I don't and can't have all the answers. I am laying down my need to always know what the plan is (this one is difficult for me). I am beginning to yield to the movement of God.

If I choose to believe these things, it means that God really does have everything in His hands, including me. It means that those hands are loving, competent, and undefeatable. Not only that, but He sees me, He cares about me, and He is actively equipping me to navigate what is in front of me. He is that active and attentive. It seems like a simple enough idea to grasp, until it involves actual, real-time submission. Here is where the rubber meets the road.

How do I submit to God when I don't understand what is happening? Submission in the face of the unknown feels counterintuitive. Acting and making decisions requires some form of situation assessment and understanding. How can I just relinquish my need to have answers when people are looking to me for a lifeline? Yielding means trusting that there is a plan and loosening our grip on the need for God to let us in on it. We are not bereft, though. In Scripture, through spiritual gifts and intentional pruning, He has already provided us with the answers and tools we do need to react to our circumstances. In chapter 5 of this book, I will dig deeper into what it looks like to get out of God's way and why it is important to make that a discipline.

Yielding means trusting that there is a plan and loosening our grip on the need for God to let us in on it.

Free Will and Suffering

The issue of why God allows or tolerates suffering has been debated, dissected, and unpacked thoroughly in sermons, commentaries, books, articles, and on social media. There are exhaustive resources out there, and I don't profess to be the expert on this topic. Toward the advancement of the idea of yielding to God, I submit my best working understanding of how the allowance of suffering dances with our ability to survive it with insight and refinement.

Every day, we get to wake up and choose who to be. While it may seem like impulses, habits, compulsions, and muscle memory take over many of our behaviors, we actually do have control over our choices. I would like to think that I wake up and choose good every morning. Odds are, if I don't put some energy behind it, I am not engaging in thoughts and behaviors that those around me might characterize as good. In fact, unless I am being vigilant to monitor and challenge my thoughts and behaviors, I hurt those who are unfortunate enough to dip into my orbit. And that is all it takes. I've done it. I put the very thing into the world that I spend my professional life fighting. I am a catalyst for suffering.

Why does God let me do this? Why does He allow me to make these choices? If a byproduct of free will is pain, why doesn't God take away my ability to choose? If we can agree that suffering is, in part, a consequence of human choice, then the question becomes, Why does God care so much about choice? If God is willing to tolerate and allow suffering in order to protect my ability to choose, what does that say about choice itself?

Let's circle back to what we believe about who God is. If we believe that God's plan is intricate, perfect, and complete, this dance between free will and suffering has to be part of the plan. If He created space and time to fit flawlessly together for His glory and perfect expression of love, surely our individual and collective gifts of free will are an intricate piece of the story. It is a struggle for me to look at my own pain and the suffering of the people I love and make sense of it. Is my ability to choose worth the price of the suffering I cause? The suffering I see? Is giving people choice worth the depth of the pain I have felt? Is it really worth the ultimate suffering Christ experienced on the cross? Free will costs so much. In his book, *Mere Christianity*, C.S. Lewis states:

> If a thing is free to be good it is also free to be bad. And free will is what has made evil possible. Why, then, did God

> give them free will? Because free will, though it makes evil possible, is also the only thing that makes possible any love or goodness or joy worth having.[2]

God is good. And all-powerful. And all-knowing. He sees everything, including me. He sees how my yesterdays and tomorrows intertwine with eternity. He saw these things when He breathed the universe into existence. He knows every tear I have shed and every piece of me that has shattered. He has seen the destruction my hands have wrought. And still, He lets me choose. In fact, my ability to choose must be so valuable to my God that He is willing to allow me to suffer for it.

I am not suggesting that the only purpose of suffering is to facilitate free will. I try not to put God in neatly constructed boxes like that. I am, however, recognizing that there is a definite relationship between free will and suffering. That relationship indicates there is something major at play. It points to our ability to choose with an exclamation mark! When viewed through this lens, choice itself becomes vital. Critical. Life or death, even.

This is where I am found, on my knees, engulfed by what feels like unsurvivable pain. My hands feel inadequate to hold anything securely, and yet I am still reaching for something to anchor me. If I look or listen hard enough, there is light. While I might not be able to see it clearly, there is hope and promise in the light. All I need to do is choose it. I don't even have to know what that choice means in the moment. Choose it now and choose it again in the morning. I might have to renew my choice by the hour. By the minute. Something profound happens when we actively and intentionally choose God. You are choosing to engage with that idea simply by reading this book.

These three WHY ideas will hopefully help us understand more about how to think about suffering and grief as we work through it.

Who am I amid suffering? *How* can I move through the world carrying this thing I now know? How can I *Yield* to the movement of God? The rest of this book is dedicated to dissecting these ideas further toward the ultimate goal of empowering us to choose Him, especially amid suffering and grief. I hope what you find in the following pages makes that choice easier.

Reflection Questions

1. How do you balance the needs and sufferings of the world with your own? How well do you tend to both aspects of grief, others' and yours?
2. What has surprised you about grief?
3. In what areas of your suffering are you holding tightly to control instead of trusting in God's provision?

CHAPTER 2

Reeling from Impact

Understanding Shock and What It Takes to Get Both Feet Back on the Ground

When the pretense of self-determination and control is stripped away, when the flimsy, artificial layers we construct for our protection are gone, it is our morality that stares us in the face.

Robert Lesslie

Wednesday afternoon, April 7, 2021. Sometime during the four o'clock hour. I remember the sheriff standing in front of me saying very specific, very surreal words. They are words that, before that day, only echoed vaguely in the morbid imaginings of someone trying to figure out if they could face their worst fears. Yet here they were. Not echoes, but crystalized words that meant something.

Ma'am, I'm sorry. They are gone.

All of them?

Ma'am, I'm sorry.

Some deep, solid part of me responded to the words with affirmation. Yes, this is real. Okay, Amy, this is happening. This means your feet need to start moving. Where is my husband? My children? *They are safe*. Where is my brother? *He is that way*. You have to go, Amy. Move.

One step. Two steps. And I was moving. Toward my brother. Toward his wife. Toward the reality of what was happening.

I still have a hard time cramming the incomprehensible way the world forever changed into those brief seconds, the seconds when there was nothing but the sound of the words my brain isolated from the rest of my body. There would be time for that later. When I think back on those seconds, minutes, and hours after I initially heard the words confirming the tragic murders of my parents, niece, and nephew, I recognize that underneath the noise and sirens, there was a constant low hum reminding me of the realness of what I was seeing. I clung to this hum and, at times, let it crescendo into a scream. It rooted my feet to the ground. It was an anchor that pulled me back from the constant threat of total escape through detachment.

Understanding Shock

No matter what our stories entail, we can all relate to the idea of shock—the immediate aftermath of a sharp slap that skews the lines of our lives without consent. The way I see it, shock is less about denial or detachment and more about what it feels like to have to abruptly reconfigure your entire belief system to make room for something unthinkable. It is the sudden awareness that all we know and hold true has been catapulted into prolonged suspension, outside of rational space.

We can also look at it from a clinical standpoint. Shock can be referred to as post-traumatic stress. It happens in an instant and can be characterized by dissociation, blurred or hazy thinking, volatile emotion, confusion, hyperfocus or fixation, a surge of adrenaline, dizziness, or nausea.[1] For those of us who have been blindsided by sudden and life-altering trauma or loss, this clinical description might seem like an inadequate snapshot of what it feels like to have the earth fall away under your feet.

It is easy enough to point to someone on the inside of a traumatic event and identify them as "in shock." They might seem dazed and slow. They might look dumbfounded, detached from understanding. Their behavior might not make sense to you given the circumstances. The biggest hallmark might look like a disconnect or incongruence between their emotional response and the reality of the situation. For example, when someone appears devoid of emotion after learning that their loved one was in a tragic car accident, we might knowingly confirm to one another that they are in shock. And we are not far off. Our systems are, in fact, reeling from impact and searching for footing.

The experience of shock looks different from the inside. It is not an absence of emotion or comprehension—it is an overload of it. Shock, or dissociation, can happen when there is too much to make sense of, or too much to give space to, in any one moment. As a therapist who works with those who have experienced trauma, I can attest to the psychological necessity of dissociation in the moments during and after a traumatic event. Shock is not an absence of emotion or comprehension—it is an overload of it.

Shock is not an absence of emotion or comprehension—it is an overload of it.

I explain it to my clients this way. In any given moment, we have a triune awareness of our experience. In ordinary circumstances, we should theoretically be able to account for or define what we are thinking (thoughts), feeling (emotions), and what our body is doing (physical awareness). Simply put, our thoughts, emotions, and body come together to crystallize our experience of any given moment. When all

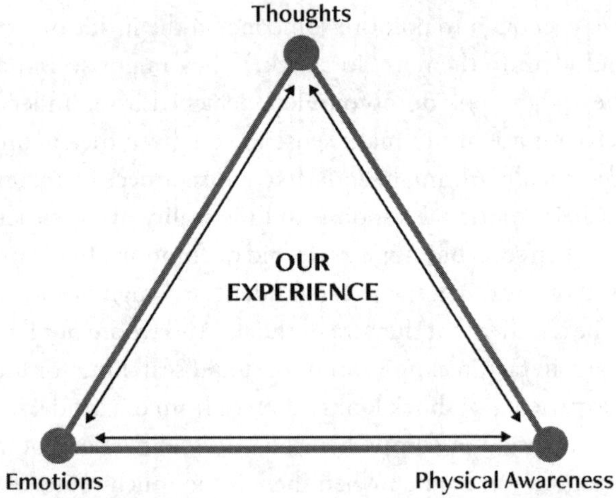

three of these things are united, or connected, this defines our experience. Memories are made of what we see and hear, what happens around us, and how we interpret and feel about those things. Memories are also made from our experience of these collective things.

When we encounter trauma, complete awareness of what our thoughts, emotions, and body are doing can be too much to handle. We might experience system overload and completely shut down. Our brains have devised a system of coping with this overload through the mechanism of dissociation. In essence, the triangle disconnects from itself so we can survive an unthinkable moment.

Let me share an example of how dissociation can be an adaptable and functional method of coping with high stress situations. When my daughter, Caris, was around five years old, we were having a picnic with my mother and her two German wirehaired pointers. Suddenly, the dogs began fighting with each other near my daughter. Caris had never had a scary encounter with an animal and so she decided

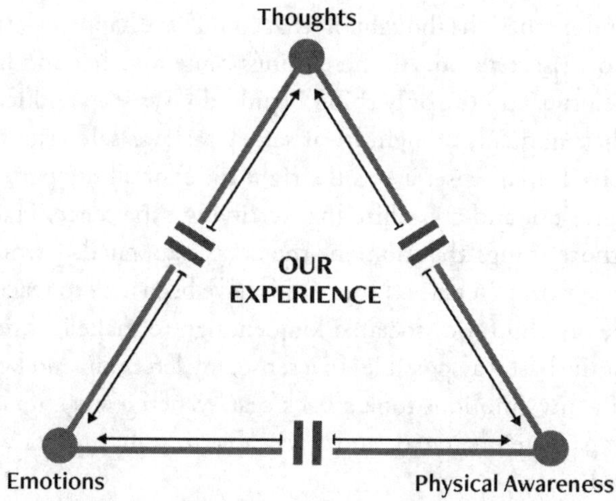

she should intervene. She happily admonished them to "stop that" as she approached the fighting animals, little hands outstretched. In the span of probably a second, I assessed the situation, realized I was not going to get to Caris in time to prevent her from contacting the dogs, grabbed a nearby towel, raced to her, and hooked my arm around her waist. I got to her just in time to pull her precious little hand out of the mouth of one of the dogs. Calmly and matter-of-factly, I wrapped her hand in the towel, strapped her into the car seat, and took her to urgent care to get stitched up. Several stiches and some ice cream later, I snuggled her on the couch to rest and watch a movie. She was fine. Content, even.

I remember going to my room when she was settled and closing the door behind me. That's when it hit, the fear and adrenaline. Everything converged at once and literally brought me to my knees: the visual of my precious daughter's little hand in the mouth of that animal, the violent sounds of the fighting dogs, Caris having to get digital nerve blocks and stitches, the thought of what could have happened, the physical toll of crisis tension. All these things came together and hit me at the same time, and the only thing I could do was stop and feel it all.

In that moment, though, I was safe. Caris was safe. The situation was resolved. In a sense, it was the right time for all my parts to converge, agree on, and crystallize that terrifying experience. Had I fully felt all those things the moment the crisis happened, I would have likely been frozen in place. I might not have been able to react in time to rescue my child. I dissociated long enough to make it through the moment the best way possible. In a sense, my left brain and body took over while my emotions took a back seat. When it was safe to do so, all three parts reintegrated, and I was able to fully process what had happened.

We might recognize this kind of dissociation in our own experiences. We willingly give ourselves over to emotion while watching movies, for example. We might let fear, sadness, and joy wash over us without questioning the plot line and realism. Our bodies and brains can take a back seat so we can give ourselves over to the emotional journey offered to us. I am definitely not a sky diver, but I assume sky divers leave their thoughts and feelings in the plane for a bit while letting the physical sensations of flight and adrenaline take over. Consider what part you might need most when exiting a burning building. I don't need to focus on the fear or the pain in my chest from the smoke. I just need my brain to tell me how to get to the exit.

These are examples of functional dissociation and reintegration. If we think of crisis response in these terms, we can recognize signs of it in ourselves and others. Sometimes this might be what is happening under the shock generated by crisis exposure. We need to survive the moment until there is enough physical, mental, and emotional safety to put all the pieces together and feel it all.

What happens, though, when there is no safe space to reintegrate? In many situations, there is no tidy resolution. When I closed the door to my room and really felt the fullness of what had happened to Caris, she was fine. I was fine. We were safe. It was neatly resolved. Unfortunately, crisis events often do not get that kind of ending. Some seemingly don't end at all. We might be in abusive situations where there is no protective space to allow vulnerability. The crisis might appear to not have an end in sight, as in the case of the COVID pandemic. We might find ourselves in a trauma situation that breeds new trauma, as is often the case with sexual assault. In these situations, the reintegration or converging of our mental, emotional, and bodily experience of the situation can take time—days or years, even. Sometimes we do not reintegrate at all.

I conceptualize the genesis of post-traumatic stress disorder (PTSD) in these terms. When our systems do not have the context of safety or resolution, our perception of the event can remain disconnected or dissociated. One or more of our parts essentially stays in a space of perceived threat or crisis. Consider what you might think of as typical PTSD-related behaviors or responses, such as hypervigilance, flashbacks, or anxiety. We can experience thoughts, emotions, and physical sensations from past crises in real time when triggered.[2] When our parts freeze in a state of dissociation, when they don't have the chance to reintegrate, they do not get the message that the crisis is over. When this happens, the part, or parts, that are still holding on to that trauma get triggered when confronted by *echoes* of danger.

My daughter, Caris, who heard the gun shots from the mass shooting, still has a very hard time with fireworks. Her brain knows she is safe, but her body goes right back to April 7, 2021. My sister, Lori, has a similar reaction to the sound of helicopters. Cognitively, she recognizes that she is far from the relentless media copters that circled our family property after the murders, but her emotions take her right back to those moments when she hears one. My startle reflex remains hyper-aroused. I often jump when my husband simply walks into a room (bless his heart). As trauma expert Bessel van der Kolk writes, our "body keeps the score."[3] Trauma imprints on our parts and they remember.

Reintegration of our parts is like three friends meeting up to tell their sides of a singular lived story. All parts are valid and important and all come together to agree on what happened. This is what I was thinking, this is what I was feeling, and this is what my body was doing. Our post-crisis job, in a sense, is to allow this formulation to happen so all three parts can confirm the experience and then agree that we are no longer in the middle of the crisis event. Even if the trauma has lasting impact, the immediate event can be separated from the aftermath, defined and described, and then processed.

This can happen immediately, or it can take time. It can be a simple, automatic matter or complicated and nuanced. I am a verbal processor, so I rely on friends and family (bless their fortitude) to help me sift through my confusion. I also have my own therapist who helps me untangle and realign. Getting my thoughts, emotions, and physical reactions out through language helps me make sense of my experience. Giving yourself space and permission to figure out the shape of your experience is an integral part of moving forward into the newness of life after crisis.

Giving Grace

That might feel like a lot of psychoeducation, but the concept is important to consider when understanding what is happening or has happened to us and those around us in the face of suffering. I think of it as our systems trying to mitigate the fallenness of the world. Our brains are trying to lessen the impact of the unthinkable. This is not how it is supposed to be. We were not created to hurt and be hurt. We were made to be in holy communion with God's creation, not to navigate its monstrous corruption by sin and fallenness. Our brains are doing their best with what they've got. This brings us to a key mechanism: grace.

Perhaps the most important ingredient in the process of reintegration is grace. Grace for self. Grace for others. No matter what we call it—shock, dissociation, trauma response—it functions to protect us. It helps us survive. It can look different in each of us and is a necessary component of how we move through traumatic situations. We need to realize that our physical, emotional, and cognitive systems are not operating in their usual modus operandi.

We were made to be in holy communion with God's creation, not to navigate its monstrous corruption by sin and fallenness.

Our tendency in the aftermath of trauma is to want to rush to be okay. And we want those around us to join us in that space, or at least to follow our lead on how to get there. Giving grace here is like telling ourselves and those around us that, even though I don't understand why you are making the choices you are making, or where you are in this grieving process, I give you permission to be there. Not only do I give you permission, but I trust you, I will get out of your way, and I'll be right beside you when you need me. Grace creates space—space to gather the resources we need to equip us for the next step in this journey.

I need that space. Through my response to my own grief and suffering, my system is doing something it wasn't designed to do. I need time and permission to move at a speed and in a way that feels survivable. And I want those around me to have that space too. In the days after the murders of my family, two of my beloved colleagues gave me a book: *Every Moment Holy, Volume 2*, by Douglas Kaine McKelvey and Ned Bustard. I remember sitting alone, waiting for the first of many family meetings to commence, and letting the book fall open in my lap. I happened upon a section on how everyone grieves differently. At the time, I did not think it was applicable to my family. Certainly we would stay in the same grief space. Of course we would be aligned in our reactions to what was happening in and around us! I could not imagine what could possibly pull us into separate experiences of pain after being bound so thoroughly together by life and then this particular death. But, at the nudging of the Holy Spirit, when my siblings and I came together a few minutes later, we read and prayed over the idea of giving one another grace in this process, recognizing that our needs might diverge. I think back to that moment as poignant and divine preparation.

Grief is not easy. The impact of loss can strip away the buffers we typically use to cushion our normal interactions with the world around us—buffers like busyness, distraction, small talk, or frivolity. Behaviors or activities that used to fill the gaps in our days might no longer feel possible to engage. My brother, Jeff, said that when someone is squeezed, the truth of who they are comes out. In the process of dealing with suffering and pain, we are sometimes reduced to acting on behalf of our most base fears and insecurities. These are the spaces where we are most vulnerable to attack (physical, relational, and spiritual) and isolation. This is also where the balm of grace can become a bridge.

We do not have to understand how someone's brain is processing their suffering in order to offer them support. It is not helpful to hold up our own behaviors and reactions to grief as the marker for how it can or should be done. One of our most important jobs as grievers or helpers is to hold the process gently and with an abundance of mercy.

Grief is not pretty. It is messy at best. Understanding a little of how our subconscious brains instantly take over our frontline defenses to protect us is a step toward being able to apply grace. As I so often say to clients in my office, *You are doing a hard thing.* It takes time and space to facilitate the reintegration of our minds, bodies, and emotions when we have experienced an earth-shaking undoing. That process requires time, vulnerability, and often the assistance of a professional. For now, though, it is enough to know that when grief and suffering unexpectedly collide with your life, just trying to get your feet back under you is a sufficient start.

Reflection Questions

1. Have you ever felt your system (thoughts, emotions, body) disconnect from itself in a moment of crisis or trauma? If so, how did that experience affect you?
2. How might your expectations of yourself and others in grief be getting in the way of grace?
3. What positive attributes have you noticed in yourself or others in the immediate face of trauma or suffering? What are some negatives that stand out?

CHAPTER 3

The Vessel

Why Your Worldview Is So Important

Listen to me, you who pursue righteousness, you who seek the LORD: look to the rock from which you were hewn, and to the quarry from which you were dug.

Isaiah 51:1 ESV

During my training as a counselor, I began to understand and appreciate the impact someone's worldview has on the way they move through grief or trauma. Prior to that, I saw grief and trauma responses as relatively steady and somewhat predictable processes that could be summarized with bullet points or a to-do list. That feels neat and tidy and safe. There are plenty of books, quizzes, and memes that will gladly identify which stage of what grief process you are in and why. It is comforting to think that if I am in the anger stage—that means I am making measurable progress! It would be nice if it were this simple. The more I journey with clients through grief and trauma, the more I realize that our response to suffering is like water. It gets its shape from its container. That container is our worldview. It is the vessel that holds thought, perception, and meaning making.

There are some great books written about the topic of worldview, so I won't go too deep unpacking the concept. But a working definition is important for the sake of making the connection between the importance of worldview and our experiences with grief. Simply put, worldview is a collection of beliefs about life that form the basis from which we pull meaning.[1] It's a set of assumptions that form the lens through which we make sense of our environment or view the world. It determines the language we use to interpret our experiences and the experiences of others.

Our Worldview as a Vessel

In his book *The Universe Next Door,* James W. Sire explores the components of several major worldviews, such as Christian theism, deism, postmodernism, and nihilism. He posits that several main questions need to be answered before we can articulate a worldview. These questions seek to explore the nature of reality, humanity, goodness, and our place in the swirl of all things existential. I recommend Dr. Sire's book as a resource if you are interested in diving deeper into the theory and differences in these major worldviews.[2] For us now, though, the most relevant consideration is why we need to conceptualize our worldview as a vessel that cradles our ideas about our life journey, and, specifically, grief and loss.

As a therapist it is vital for me to establish why my clients come to see me. Not just why, but *why*. Why does it even make a difference? Why does feeling better or being better even matter? Is there such a thing as good, am I interested in engaging with it, and do I consider moving in that direction to be important? What is intrinsically valuable about joy or even living? Not everyone can readily answer these questions, and the answers are incredibly important. As a Christian, I am committed to the belief in a present, active, and loving God who provided a perfect model of goodness and sacrifice in the form of His Son, Jesus. I can look to the tenets of my faith for an understanding of how I relate to the world around me. I curate my thoughts and live (or try to) according to the ideas that I find in Scripture. Our worldview is a vessel that cradles our ideas about our life journey, and, specifically, grief and loss.

I pour my experiences into the Christian vessel that is my worldview and allow that vessel to define the way I interpret what is happening in and around me. This process is essential when I am trying to make a decision about my relationships, my behaviors, my career, and my time. My worldview determines the nuances of my daily life as well as big picture ideas and trajectories. This is especially true when life doesn't go my way, or when I am inundated with pain. When it comes to suffering, grief, loss, and trauma, we need to be able to rest on something solid in order to make sense of things. And making sense of things often becomes our primary need when facing grief, loss, or trauma. We need to have a determined measurement of value so we can understand exactly what has been lost. Recognizing what measures up as valuable to us might be the only way to understand why this hurts so very much.

Our worldview is a vessel that cradles our ideas about our life journey, and, specifically, grief and loss.

When I tragically lost my parents, niece, and nephew in the mass shooting, it was impossible to even consider the magnitude of the loss. There was an extended period of time during which my family could only address what was directly in front of us at the moment. Being proactive did not seem possible; there was only reaction. My sister-in-law,

Katie, described the slow revelation of the losses in the weeks and months after losing her children as a "thousand tiny deaths." Every day seemingly brings with it the recognition of another subtle loss, another thing to grieve, another emptiness. Those of you who have lost a loved one will understand the slow and gradual shape that grief declares. There are the big things like not being able to talk to or hug someone again. And then there are the endless other things that also get taken away from us. Instead of an olfactory reminder of steadiness, the occasional smell of my dad's cologne brings me to my knees. The sugary snacks in my pantry aren't quite as sweet if I can't fill my neice's and nephew's pockets with them before sending them running home. Who is going to sneak mushrooms into all the casseroles my husband eats, if not my mom? Who is going to call and check in with me every Wednesday morning, if not my dad? My dad was, among other things, a piper. As of this writing, I have yet to be able to listen to bagpipe music. The wind blows differently now. The benchmarks of time's passage are steeped in recognition of what was or could have been. *Overwhelming* is too gentle a word for the recognition of those losses.

Overwhelming. It's just not enough of a descriptor. Confusing. Painful. Foggy. Exhausting. Heavy. Consuming. Dark. Jarring. Violent. Lonely. *Relentless*. Grief can be so relentless. This is why we need a vessel. My hands are not enough to hold it all. My brain feels too soft and small in the face of this reality tornado. I need something to hold this cluster of complications because I can't. And this is something too personal to ask of someone else.

This brings us to the idea of our own personal worldview. What is your vessel? And where will it take you? When, if ever, have you spent time giving shape to what you really believe? Have you checked your worldview lately for unwanted cultural influence or cynicism? For example, in my line of work, I come up against the vestiges of nihilism all the time. It can be all too easy to convince ourselves that *nothing* really matters and that at the end of all of this, there is nothing. Honestly, I get it. Sometimes *nothing* is the closest thing to relief we can imagine. Sometimes we are so exhausted that *nothing* is all we can muster. In a world of pain, disappointment, betrayal, and deep grief, *nothing* feels like a viable answer. Nihilism is absolutely a worldview we can choose for ourselves, or even find ourselves sliding into because of complex, sustained rejection or discouragement. We don't have to work too hard to even make it make sense. But does it work?

I will challenge the idea of *nothing* at every turn both in my professional and personal life. If your vessel is nothing, it is not a vessel at all. It is what it claims to hold—nothing. And it will guide you to the same place. The movie *The Neverending Story* (1984) is one of those perfect, transcendent '80s movies. It is as beloved to my kids as it is to my siblings and me. While I am not claiming that the movie is absolutely about nihilism (though perhaps it is), it still fits. In the film, the world of dreams and imagination is slowly being consumed by a rolling, dark cloud-monster called "The Nothing." It simply spreads and destroys everything. There is only one place where it ends. There is only one outcome. All roads named "nothing" lead to darkness. Worse than darkness, nothing. The argument that decency or joy can coexist with nothing can't hold up. There cannot be both.

Christian Hope

Enter hope. In *The Neverending Story*, hope is ultimately represented by a grain of sand holding what's left of imagination and curiosity. This is the seed of *something*. Hope is the place where meaning digs into the soil and takes root. It is what anchors and guides us. Hope is the seed we cling to when surrounded by the kind of nothing that grief pours on our lives. Hope. How we define and characterize hope is a fundamental declaration of who we choose to be. And this is where our worldviews can take us on different paths.

Christian hope is no stranger to theoretical dissection. There is enough commentary on Christian hope to keep someone busy for a lifetime. Right now, I am not interested in what the commentaries have to say. I am more curious about my own experience of what hope actually looks like in practice. I am a Christian woman whose life was shattered by tragedy. What did Christian hope look like for me then, when depravity itself broke its way into my family's most sacred places?

Worldly hope is akin to a wish. I hope or wish my desire will be fulfilled and delivered in precisely the way that I have decided is appropriate. Christian hope is about submission and promise. Promise that God is who He says He is. Promise that it is all true—everything. He knit me together in my mother's womb. He will instruct, and teach, and guide me. He will never forsake me. He is merciful, and steadfast, and good. His love is perfect. His love is perfect. *His love is perfect.* This is my vessel. Unlike with earthly hope, my desires take a back seat to the simple truth of who God is. I lay it all down at the feet of the God who sees me. This is the ship I found myself in when the world around me was blindingly uncertain. My feet hit solid ground. I was able to stand. And walk. And act. In the moment that lasted a lifetime, I was reduced to the only thing that I knew to be true. God is who He says He is. God is good. And He is holding this together.

Christian hope is about submission and promise.

In that, and the moments after, I also had a slow realization that I did not need the answer to the question *Why?* Why them, why us, why my family, why this way? The *why* became less important than the *what*, which was that God has got this. It is what it is. I don't need the answers right now, because I know who the author of the story is. I can wait. Part of my hope is the promise that my next chapter, the eternal chapter, will carry whatever answers I really need. In the meantime, God continues to instruct, guide, and teach me. I am okay with that. More than okay with that, I am here for it.

Hope looks like different things to different people, though, and might be harder to pin down if you don't come from a background that emphasizes it. Hope may feel to some like more of a wish or an intention. At the end of hope, though, there is always meaning or value. I hope in what I hold dear. For those of you wondering what your vessel is made of, ask yourself what truly matters, what you find yourself hoping for. Goodness? Faith? Love? Connection? Money? Success? Find what matters to you, go deeper, and give it shape. Take the idea of what really matters and follow it to its beginning. *Does it coincide with what I believe to be good? Where does goodness come from and where does it end? What is the genesis of what matters?* My heart hopes you find Christ in that journey.

Our Identity

It is these ideas that hold us when we are buffeted by storms. We need to answer these questions for ourselves now, before the storms hit, if they are currently swirling around us, or if we are assessing the damage after the storms have settled. Again, our response to grief, loss, tragedy, trauma, and unrest is like water. It takes the shape of the vessel it is in. Or, in the case when there is no vessel, it just falls away from us. Like Robert Smith from The Cure writes, "And every time I try to pick it up like falling sand, as fast as I pick it up it runs away through my clutching hands."[3] In my experience, that is the hardest reality of all: having these important pieces to make sense of, but no way to carry them. No place to gently, and with purpose, lay them down.

So we consider the bedrock that is identity. Erik Erikson, the developer of psychosocial theory, divided the human lifespan into eight stages, each with its own mastery to achieve.[4] For example, infants have the job of developing fundamental trust versus mistrust based on how their needs are met. If they have unresponsive or neglectful caregivers, if their basic needs of safety—warmth, food, and human contact—are not met, infants learn that the person assigned to them cannot be trusted. Conversely, if we tend to infants with care and responsiveness, they develop a sense of hope that the people around them will continue to provide safety. Note that hope, in this sense, is more like confidence than a wish. I don't think that it is a coincidence that we engage with a Christian-like hope even as infants, and that it is a fundamental part of functional development.

Let's jump past a few stages to the one that is most relevant to our topic. The development of identity (versus role confusion) falls to the teenager and young adult. Developmental psychologists identify this period of growth as the appropriate time to internalize morality and nurture a fidelity to values.[5] Simply put, this is the time for us to figure out who we are, what we stand for, and why. In a world where definitive identity can be eschewed because it is deemed too limiting or exclusive, developing this sense of *knowing* who you are can be an uphill battle. Culture can give us mixed messages about this process. The world seems to be telling us that no one else can tell us who we are, while encouraging us to keep our identity open to outside influences.

Besides cultural interruptions to identity formation, other roadblocks can hinder this process: trauma, housing disruption, family separation, illness, or abuse, to name a few. We might not make it to age 25 knowing all we need to know about who we are. For most of us, it is a lifelong journey. God doesn't finish with us when we graduate to the psychosocial stage of adulthood. Praise Him for that! The important idea here is that grappling with the *what* and *why* of identity during our formative years, and after, leads to a deep and essential commitment to morality and values. In other words, it is the basis for worldview. If you feel that your search for identity was impeded in your developmental years, don't worry. Taking the time to define your worldview is incredibly valuable at any stage.

We can all benefit from taking a pause here to consider what we know about what we know. While I can absolutely articulate what I believe and what my worldview is, the authenticity of it is measured by how congruent it is with my actions. In *The Universe Next Door*, James W. Sire says that "our own worldview may not be what we think it is. It is rather what we show it to be by our words and actions." His advice for challenging the authenticity of our worldview is to "reflect and profoundly consider how we actually behave."[6]

This insight ought to be touchy for all of us. We are all capable of straying from even our most core beliefs. Even scarier, we are capable of engaging in the kind of mental gymnastics, or cognitive distortions, that convince us that discordant behaviors are justifiable. Oh, the damage we do to others when this process goes unchallenged. This is the condition of mankind. We cannot separate ourselves from our fallenness. Living congruently with our values takes intention and discipline. It also takes courage to be able to acknowledge discrepancies, confront the truth of our beliefs, adjust them if necessary, and correct the course of our behaviors. It takes courage, and probably sacrifice, to come into alignment with who we feel we are supposed to be. That is certainly a Christian view of growth and identity. The key is to take time to confront this idea and turn insight into change with action intended to increase congruency.

You have already determined that you are important by picking up this book. Your pain and the journey you are on is worth confronting. Not only that, but *who you are* is valuable. Your thoughts, beliefs, and words are valuable. No one can determine them but you. Again, your choice matters. It determines how you walk through the fire and matters deeply to God. Figure out what is at the core of your identity. Seek it now. Find it. Don't stop until you have clarity about what the world is and who you are in it. Finally, living those truths out loud will give them integrity and be the thing that catches you when you need it most.

Reflection Questions

1. When you consider your worldview, what comes to mind? Is it hazy or unclear? Is it yours, borrowed, or forced? Does it feel strong enough to stand on?
2. What kind of questions do you need to ask yourself to dig deeper into your worldview?
3. How are your thoughts and actions in grief lining up with your values? How are they incongruent?

CHAPTER 4

Help

Why It Is Vital to Invest in Your Relationships Now

Community, dear friends, is the place of intimacy, the place of fruitfulness, the place of celebration. I want to say that for you, for all of us, that is the place where the vision of hope is being born.

Henri Nouwen

My father, Robert Lesslie, was a physician and an author, among other things. If I've ever met a true renaissance man, he was it. He spent his adult life caring for peoples' physical needs in his clinic and spiritual needs through his faith, relationships, church involvement, articles, books, and volunteer services. He saw the fragility of life first-hand in the emergency department and urgent care clinic and wrote about it often. He spoke at length about the need to show the ones you love how you feel. *Treat them well. Love them. Life is fragile. We are not promised tomorrow*. Now that he is gone, his own life so abruptly cut short, it is an amazing blessing to have his words on this idea so clearly and prolifically spoken. I urge you to take the time to peruse his writings to see what he has to say about the brevity of life and what that implies. Now, I want to expand on that idea by highlighting another reason to cherish the ones you love.

I live out in the country, and the property that belongs to my husband

and me is attached to a bigger piece of family land. My parents' home was here, as is my brother Jeff's. In its glory, the family land consisted of about 50 acres of peace, beauty, and adventure. My three children grew up running through the woods and over the pond to get to Nana and Papa's. Holidays out here were exciting and full. Our homes overflowed with family, noise, laughter, and love. The paths between our homes were well-worn and well-loved. My dad was a visionary who saw what he wanted for his adult family and made it happen. He built a giant invitation for his kids and grandkids to join him and my mother in life. He lived what he preached, and what a blessing that is. The land was a space my parents set apart for quality time, fellowship, and meaning. I can confidently, and with eternal gratitude, say that it fulfilled that purpose.

On April 7, 2021, our refuge was infiltrated by unspeakable violence. In the immediate aftermath of the shooting, it was crawling with law enforcement, first responders, and related service members. They were everywhere—on the roads, in the woods, in the air. As a result of this event, the occupation of our home was instant and jarringly abrupt, although entirely appropriate and handled with the utmost respect. It's fair to say that there was an element of powerlessness and a sense of bewilderment as we watched the legal response to the tragedy unfold in front of us. In a short amount of time, however, the influx of law enforcement and media began to give way to another wave of intervention. This wave came in the form of our people. As news spread and the impact of the tragedy was felt, our people began to mobilize. I can't write this without getting emotional.

During those first days, a handful of immediate family and close friends were consistently on the property. I remember having no idea what I needed but being aware that my needs were being covered. The well-worn paths between our homes were supporting an entirely different call to connection and facilitation. There is no question that being that close to the tragedy was surreal and emotional. Something else was happening at the same time. Some of those beloved helpers have confirmed that I am not the only one who felt the movement of God on the land. I believe the close friends and family who were there can attest to what it felt like to walk those grounds. You might imagine that there was a blanket of horror that coated everything. But from my vantage point it was *sacred.* Not just sacred, it was *holy*. Not just holy, it was *miraculous*. Of course, the shock and grief continuously brought us to our knees. Our suffering at that time was sharp, breath-stealing, and acute. To say we were overwhelmed by pain and loss is sadly understated. But something else was also happening. There were multiple truths intertwining and overlapping with the pain. We were being fed and *protected.*

The paradox of feeling protected after such a violating assault is a strange thing to live through, maybe tougher to explain. The first layer of protection we felt was spiritual. We were gifted an immediate and lasting awareness of peace and connectedness. The veil separating us from the heavens was barely-there thin. We were surrounded by the miraculous. I have described it as a spiritual dome of protection over the whole land. I mean that literally. There was undeniably a holy witness to our pain. Again, this is where worldview wraps itself around you. I know what God feels and sounds like. I recognized His presence when I needed it most. He wasn't a breath away; He was there, in every blade of grass, every brush of wind, every step we took. He did that for my family then, and He hasn't stopped.

I want to take a moment here to acknowledge that the feeling of literally being covered by God in a grief or crisis situation is not everyone's experience, and it is certainly not contingent on the strength of one's faith, deservedness, or magnitude of loss. While suffering is not a unique experience, the shape it takes in our lives and our minds certainly is. How God chooses to reveal Himself to us in our moments of joy and pain remains a mystery, but His attention and concern for us is not. If you are reading this, wondering where your dome of protection is, you are not alone. You are not alone in your wondering or in your pain. Your suffering, too, has a holy witness. To address the question of God's felt presence, I would simply ask you to trust in the nature of the God who sees all of your days and is working for your good, whether you can perceive it or not. Pray for Him to reveal Himself to you as the great comforter and provider and wait with faith. Your suffering has a holy witness.

The second aspect of protection that covered us so completely was relational. This is what I really want to unpack. I grew up fiercely clinging to my family identity and the relationships within. It seems this is how I have been since the womb. This is not to say my family of origin hasn't had our disconnects and discord. We are far from perfect. We have all the black sheep (of which I am one), personality issues, stubbornness, and dysfunction of any typical family. The Lesslie clan motto is "grip fast," however, and boy do we. We grip fast to our ideas, our dreams, our people, and one another. We hold on tight, despite, or maybe to spite, the storms.

Your suffering has a holy witness.

Cultivate Your Community

The idea of investing in or holding on tightly to those you value extends beyond family and into friendships and community. We are not promised that our tomorrow will look like our today. When my family's reality was devastatingly altered, the relationships we had built as individual families and those built by my parents, niece, and nephew surrounded us with powerful provision and protection. This idea cannot be overstated: *Consider and cultivate your community.*

Scripture offers a great blueprint for how we can conceptualize the idea of cultivating community. Christ exemplified a perfect model of relationships. Imagine concentric circles representing Christ's earthly relationships. Before the resurrection, the Holy Spirit and God were at the center of Christ's concentric circles. We can be the same. At our center there is us, and there is our relationship with God, if we allow it and prioritize it. This center space is the most important. Henri Nouwen refers to it in terms of personal intimacy, saying, "The place

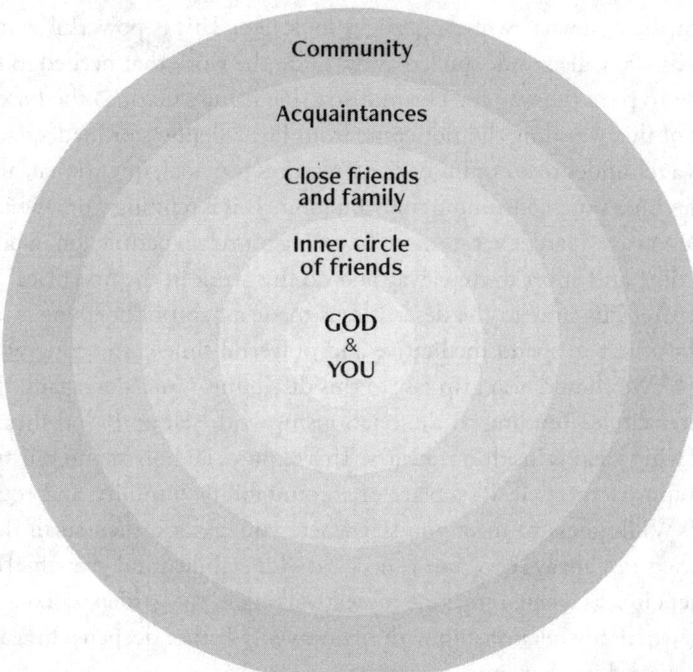

where you can hear the voice of the first love is called solitude. Solitude is where you can listen and where you can hear a voice that says, 'Do not be afraid. I love you.'"[1] It is the core of who we are and the source from which all other relational characteristics or impulses flow outward.

God and You

Before He began His adult ministry, Jesus spent 40 days in the desert, wrestling with temptation.[2] Like so much of Christ's ministry, this is a model for us. There is something powerful and often neglected in the idea of what it might look like for us to give ourselves permission to tend to our personal needs in this way. I often talk to my clients about what their "desert" wrestling might look like. This is powerful work. Christ was called aside, pulled away, to do the work that needed to be done to pave the way for His ministry. He is fully God, so the necessity of this wrestling did not come from His fallenness or inadequacy. It is a reminder that our humanness requires personal, intentional, and sometimes time-consuming confrontation. This is refining work, which is never easy. Harder yet seems to be giving ourselves permission to take the time and effort to step away and do this work in the first place.

After His time in the desert, Jesus made a habit of breaking away from others to spend meditative and prayerful time connecting with God.[3] We should also grip fast to this discipline. Consider, again, the center circle. Tending to the relationship with self and God means knowing what is inside our hearts. This requires an honest and nurturing approach to self. It's a place of accountability, humility, and courage. Willingness to meet yourself where you are and then sit in that space is not always easy, but is necessary for stability and growth. The inner circle is where I imagine myself walking in the garden with God. It is where my relationship with my heavenly Father deepens. I mean, if Jesus did it, count me in.

> Willingness to meet yourself where you are and then sit in that space is not always easy, but is necessary for stability and growth.

It is worth it to address the stigma of selfishness here. The idea that we might be considered selfish if we step back and devote resources to confronting the self, assessing our weaknesses, and regrouping is the main deterrent I encounter with my clients. Why is it so hard for us to prioritize emotional health and wellness? If you struggle with this, I recommend talking to a trusted adviser to help you sort through the difference between being selfish and being a good steward of the vessel God has entrusted to you. Sometimes a sabbatical is exactly what is needed to prepare us for what God has in store.

Inner Circle of Friends

The next ring in our concentric circle is our closest friends. This is where we begin to look at the depth of what community really means. For Jesus, it seems to have been James, John, and Peter. For most of us, there is only one person, or a couple, or maybe a handful, of people in here. The number of people is less important than the quality of the relationships. And guess what? *We get to choose them.*

It is worth saying that we are all out here waving red flags. All of us have the capacity to hurt others and sink relationships. We disappoint. Hopefully, we are tending to our own depravities in our inner-circle wrestling. Despite our shortcomings, we generally think of ourselves as capable of friendship and think we can easily recognize bad apples. As a therapist, though, I often find myself talking about the concept of safe people. It is not something we are necessarily taught as we are growing up, and it shows in our adult relationships. There are unsafe people in all aspects of our lives: workplace, school, neighborhoods, church, and certainly within our families. Recognizing them and setting appropriate boundaries is an important discipline.

There is a list of behaviors that are identifiably unsafe for intimate relationships: emotional volatility, manipulation, deceit, controlling tendencies, and physical and emotional violence, to name a few. I am talking about abuse. We can't always disentangle ourselves from abusive people, but we can try to set boundaries that limit their access to us. Abuse comes in many forms: physical, sexual, emotional, financial, and organizational. People who engage in cruelty, coercion, and manipulation cause damage. People who *require* you to be unhealthy to be in a relationship with them cause damage. People who refuse to be accountable for their behaviors cause damage. When choosing your closest friends, these are qualities to be mindful of. These people have intimate access to you. Make sure they can be trusted with your vulnerability.

Even more important than seeing the red flags, though, is recognizing and investing in the people who are reliably safe. When assessing this kind of reliable safety, specific relationship roles will determine what characteristics of safety we need to focus on. For example, your pastor does not need to be your go-to person for medical advice. They probably do need to be a consistent shepherd and actively accessible. Your therapist should be the person with whom you process your innermost struggles, but they should not allow you to join them on vacation (see the movie *What About Bob?*). Your coworkers can support your career victories and deficits but likely don't need to know the ins and outs of your financial portfolio. Your adult children might not need to be the people with whom you share your secrets, but there is infinite value in being able to share your life with them. In short, ask yourself, *Are my expectations appropriate according to relationship roles, and are my people holding their relationship jobs reliably and with integrity?*

When it comes to your inner circle, the answer needs to be an absolute yes (amen!). These might be the people who will jump in the car to get to you because they know what you need before you can articulate it. They might be the ones you trust to run your household when you just can't. These might be the people who can sense the subtle changes in your voice that are your barometer for wellness. They do need to be people you can trust, because when the sky falls, they will be the ones holding the umbrella.

I will reiterate that while the people in your inner circle can certainly be family members, they might not be. This is okay. We do not all have family members who are trusted friends. All of our families are fallen, but some of our family relationships are too dysfunctional to be considered safe. Relational boundaries are scriptural.[4] You have permission to be healthy! I do think it is important to note that boundaries are created and maintained in order to nurture healthy relationships with others, not to shut them out. Boundaries are not barriers; they are appropriate limits that allow for safe connection. Boundaries are doors that restrict unhealthy access so the windows of true connection can open freely.

While those in our inner circle are not perfect, and they will make mistakes that hurt us, they are set apart. In theory, our closest relationships are safe, reliable, honest, accountable, and capable of putting in the work. My friend Courtney calls these people her family of choice. We are gifted with these people along the way. Some stay for a bit, others a bunch. These are people who are willing to walk into hard conversations with you. They are willing to tell you hard truths and hear them from you. These are the people who figure out how to recognize, support, and accommodate your growth. Who have been these people for you in the past? Who are they now? Who do you want them to be? Finally, what do you need to do in order to solidify them as inner-circle friends?

One thing we can do to nurture our close friendships is not require our people to be Jesus. They aren't always going to meet the mark. I spend a good bit of professional time talking about the idea of being messy. It is one of my favorite aspects of being human and of earthly relationships. Messy might be fumbling through hard conversations. Messy might be saying the wrong thing. Messy might be not knowing how to show up in a situation but showing up anyway. Messy definitely includes getting it wrong even though our intentions are right. We have the capacity to be messy with one another and survive. In fact, messiness is where so much of the good stuff happens. It is where we are pushed and confronted. It is where we often expose our ugly places to ourselves and others. It is where we find grace. It is okay to be messy. Equally important is that it is also okay for your people to be messy. One of the most profound places to meet is where the human in me collides with the human in you. Dealing with messiness requires communication. Communication requires humility, honesty, love, and courage. For the source of those things, refer to the center of your concentric circle.

One of the most profound places to meet is where the human in me collides with the human in you.

Close Friends and Family

As we move outward from our inner circle, the roles of our relationships evolve. The next ring might have more people in it. These are very much still our people. They are friends and loved ones. The rest of Jesus's disciples would probably be in this group. If someone throws you a surprise birthday party, these people will be there if they can. They are laugh partners, lunch dates, shoulders to cry on, and are absolutely your pinch hitters. They are people you can call on to check your mail for you or come and get you when you have a flat tire. To quote my friend Courtney again, these people are your tribe.

My parents have a tribe. My husband has a tribe. My children have tribes. My adult siblings and their spouses have tribes. In the days after our tragedy, we began to feel the overwhelming energy of our tribes as they reacted, perceived needs, responded, and converged. From our insulated vantage point, we saw some of that movement but felt it even more. When I think back to that time, I remember perceiving an intimate but powerful dome of spiritual protection over us, and then a buffer of relational protection and provision that extended outward further than we knew.

These people not only tended to us, but also to one another in their grief and in their need. Practically, this looked like food, flowers, car rides, phone calls, errands, coordination, and event planning. What it felt like, though, was being surrounded by soft, warm, giant arms. When the sky fell, these arms shielded and held us.

My father used his writing to consistently remind people to hold loved ones near because we don't have infinite opportunities. I double down on that sentiment and add that we need to create, tend to, and deepen the relationships that matter to us because we are going to inevitably need them in the trenches of pain. All of us know the pang of isolation and loneliness. We know what it means to feel abandoned in our suffering. Some of that is a characteristic of the human condition. Some of it can be mitigated by making the effort to connect. Much of the support that came our way after the mass shooting was not from our direct contacts. Many of those relationships were bound to our story by other family, friend, and community connections. When it mattered, though, we were found by people who wanted to help, and I credit so much of that to my parents' legacy of connection.

Tend to your tribe. Know them, invest in them. If you are reading this wondering who your tribe might be, or how to get one, look around you. What kind of communities are you part of? We most often find relationship within families or at work, school, church, or in other community organizations. Experiment with what it feels like to open yourself up to the potential of deepening some of those relationships. If you need help, reach out to a counselor or someone in a trusted advisership role to help. It is worth it.

Acquaintances and Community

As the rings of our circle move outward, roles and responsibilities change. My dentist might not be in one of my inner circles, but he is definitely an important person in my model. I know who to call when I have a toothache. Likewise, I know where to go when I need work done on my car, when I need spiritual advice, when I don't know how to parent through something, and when I need someone to hold my hand as I stand at a scary crossroad. The further out the rings get, the less intimate the relationships become. The connections are still important, though. Community relationships like school connections, church families, and coworkers might fall in these categories. Friends of friends might make up some of the people in the outer circles. While they might not be your besties, they still play an important role. These are still people who will rally behind your meal calendar or fund raiser if needed. The culmination of these people is your community, and the integrity of your community is yours to assess and adjust. As your grandmother may have advised you, it matters who you surround yourself with.

We tend not to evaluate our communities until we are in dire straits, if ever. Instead, this should be a lifelong, active consideration. A tomorrow will come that is unexpected and life changing. Cultivate your community garden now.

Finally, when the time comes, ask for help. Allow people to show up for you. Let them in to do whatever they do best. Let them in with all of their messy glory. When those arms circle around, open yourself up to their warmth and simply allow it. Grief is hard enough. Let someone else fold your laundry.

Reflection Questions

1. What areas of your concentric circles need more attention? Are you open to exploring those connections and working on investing in them? If not, or if you are unsure, what is holding you back?
2. Who are the people in your innermost circle? Are you nurturing those relationships appropriately? If you can't think of who they might be, consider who you want them to be and how you might deepen those relationships.
3. What communities are you a part of that offer support to others? Are you accessible if someone else has specific needs you might be able to meet?

CHAPTER 5

Making Room for God

How to Get out of the Way

The doubt of wonderment is a normal part of a life of faith, and it's spiritually healthy when it drives you to bring your confusion to the One who has no confusion.

Paul David Tripp

Even though there were no cell phones poised to record shenanigans when I was in high school, there may be several witnesses who can attest to my adeptness at tomfoolery. My friends from back in the day know who they are and the kind of antics I am talking about! The truth is, I was passionately difficult to manage. When my past life as an angsty and rebellious teenager comes up, it's often in a lighthearted or joking way. But when I counsel teens who are struggling with fitting into the mold their parents, churches, teachers, or even friends envision for them, I truly get it. That was me, and it's not funny when you are going through it. Being different, or "extra," or "a lot" can be hard to make sense of, even if there's a brash exterior on top. A certain cluster of troublesome characteristics (talkative, bossy, contrary, etc.) can be affectionately perceived as precociousness when we are young. *Precocious* stops being cute when your parents start getting phone calls from teachers and other parents or get side-eyed glances from their peers at church.

My mom says that while my sister came out of the womb a delightful charmer, I came out with one eyebrow raised over a decided frown. She used to laugh about how Lori, my sister, would just toddle up to people and make friends while I would glare at them, sizing them up for their worthiness. When my language skills caught up with my facial expressions, I began to question the credentials of any adult authority figure placed over me. Even as a child, I needed proof that someone was worthy of my respect, or even my attention. So I challenged all the adults in my life: teachers, babysitters, church staff, neighbors, relatives. Assessing how much control adults had over a space or system came naturally to me. I wasn't content to just notice holes in authority; I poked them. I redrew as many lines as I could just to see who could keep me in a box.

I discovered that I could get away with challenging systems a lot, if not most, of the time. Now that I am an adult, I can understand that a lot of people just don't have the energy, patience, or inclination to rise to the challenge of a mouthy 12-year-old. A lot of people are just simply burned out. My poor seventh-grade science teacher just wanted to teach us about the solar system and go home. She was likely near retirement, longing to hang up her ruler with little ado. And in I walk, confident, mouthy, and sporting the sense of humor of a tween boy. When I say bless her heart, I mean it with all of mine. Most of my childhood Sunday school teachers were older volunteers who did not sign up for what I walked in with. I surely tested their aptitude for apologetics!

I get it now. I get the irritation and apathy. I get why most of them didn't want to enter into those choppy waters or challenge me back. But what I found then was something lonely. I wanted to push against the world and have it push back. Instead, I think I became aware, on some level, that a lot of adults are just winging it and don't have room for tomfoolery or an inkling of what to do with it. Figuring out how to set boundaries is difficult, and there are plenty of adults who aren't sure how to give them integrity. Occasionally, I am one of those adults. I definitely don't always know what to do or how to handle a situation. I am often tired and just want to wave a white flag. As a child, I think I needed something more, though. I wanted to feel that there were constraints in place that would not allow me to fall over the edge. The truth is, there really aren't. We can fall far and completely. That is scary.

My insistent prodding gave way to open rebellion, and I spent a lot of my teenage years breaking rules. I had plenty of coconspirators, and most of them are, thankfully, still in my life (shout out to my gritty Gen X sisters). But despite the company I had along that part of my journey, it was ultimately a lonely place to be. Again, I was looking for a solid structure in which to exist but was coming up short. Nothing was pushing back. By the time I went off to college, I was fairly lost.

I still struggle at times to make sense of those lost years. I have never been able to pinpoint any specific change that led me to a straighter path. What I do know is that at some point, I felt found. I found my footing, found momentum, found my degree. I found my husband, found my faith, and found my future waiting. I can't take credit for this. It is clear to me that God grabbed me with His shepherd's crook and pulled me back from the edge. Praise Him.

For all of us who imagine that our parenting is going to be magically different, and that we are somehow going to find the perfect parental balance, be careful where you place your bets. The background of my childhood was a solid Christian environment. I had two loving parents who were strong in their faith. They were active, successful in their individual endeavors, and beloved. They had us in church on Sunday mornings and Wednesday nights. We went to church retreats and camps. My dad was an elder and my mom was a Bible teacher. My friends were in church with me, played church basketball alongside me, went to camps with me. We did all the churchy things. And yet I still felt compelled to unravel it all to see what was on the inside. I sized up everything for its worthiness. I didn't like the inconsistencies I found and felt let down, disillusioned, and maybe even betrayed. When I challenged the world to show me what it was made of, it came up short. At that age, I couldn't understand what the disconnect was.

As a Christian adult I know why: We live in a fallen world. Sin is everywhere. The world will always come up short. Growing up, I was surrounded by people who told me that this was the case. I grew up hearing about Adam and Eve, the snake, the forbidden fruit, and sin. I heard all the words that described the plight of humankind. Despite being familiar with Christian lingo about fallenness, I needed to discover it for myself. I needed to see the battlelines between good and evil. I needed to claim my own experience in order to validate it. When I learned the word *Mine!* as a toddler, I gripped it fast. My questioning and rebellion was as complete as it was honest. I was Thomas, who needed to touch the wounds of Christ to be sure they were real.

There are two ways we can look at this. If I had gotten out of the way of God, if I had trusted the leaders in my life, I might have been spared a lot of difficulty. I might not have been hurt as much or hurt as many people. My poor parents might have gotten more sleep. My path might have been straighter and might have led me sooner to the field of counseling and ministry. If I had just gotten out of the way of God from the beginning, who knows what mysteries and moments of grace might have emerged?

I didn't, though. A lot of us don't. This brings us to the other way of looking at it. What if I get out of His way *now*? What if I take a leap and "trust in the Master I serve," as my former pastor, Barry Dagenhart, has put it? What if trusting that Master means accepting my life as it has been—in all its crookedness—and acknowledging that it might be part of what He is doing *now*? What if my wonky, rebellious, winding past has been a part of what He has been doing all along?

God's Sovereignty

For us to move in this direction, toward God's intricate and purposeful plan, we need to establish that *this* is how God works. That means digging into the issue of sovereignty. *Sovereignty* is defined as ultimate authority or supreme power. Scripture makes this idea more robust by bringing it into the now, describing it as present and active (Romans 8:28; Job 42:2). God's sovereignty is not just a distant ordering of events, but also an intimate involvement with singular, earthly moments and decisions. God moves mountains and even works in the minutia. Job 42:2 says that God's plans cannot be thwarted. Proverbs 19:21 tells us that it is the purpose of God, not man, that will stand. Ephesians 1:11 says that God "works all things together according to the counsel of his will" (ESV). Scripture repeatedly highlights God's purpose and its immutability (unchangeability). You might have heard it, and I have definitely said it: God's got this. He is sovereign. His plans will come to fruition, despite the machinations of man, despite my rebellion, and especially despite all of my good intentions.

> God's sovereignty is not just a distant ordering of events, but also an intimate involvement with singular, earthly moments and decisions.

This is a wonderfully comforting thought. God is in control of it all. He is in control of the beginning of time, the dimensions of eternity, and everything in between. This is a powerful realization to embrace and submit to. It is also complicated and difficult to absorb when you are hurting. God's sovereignty is tough to sit with when you are in the middle of the fire. If God can do all things, and if He has complete power and ultimate authority, why did it have to happen at all? Why did this have to happen to me? Wasn't there another way? Why is there so much suffering? There are plenty of brilliant theologians and lofty commentaries that unpack the issue of sovereignty more thoroughly than I can. What we can do here is try to increase awareness of it, accept it, and come to terms with God's sovereignty where it collides with our perception of it: *in our pain*.

When my grief is primal, the why of it all churns like a storm. My heart wants to negotiate a takeback. *Okay, God, I can understand You needed one of them for Your purpose. I can accept You might even need both of my parents, but all of them*? *All of them*? The words are hard to write. My human brain can come up with a lot of different versions that can surely accomplish whatever it is that God is doing here. How about a glossy miracle, even? Wouldn't that do the job instead? Did it need to go so far? Don't get me wrong, I do not believe that God *did this*. But He is sovereign. He knew it was going to happen, and He did not stop it. That is the truth of it. This truth is what sometimes violently collides with my pain.

Misunderstanding can so easily give way to anger at God, which He can handle, if I do not make room for another layer of His sovereignty. He knows how all our pasts and futures weave together for His ultimate purpose, and *He still did not stop it*. He sees it all. He sees my family. He can name our tears. This is still the portion we carry. God knows every thread that is connected to the story of my family. He intimately understands every thread that is connected to our pain and suffering, and He grieves with us. What He doesn't stop, He allows.

What He allows, He works for good. He redeems, and He redeems completely. I trust it. I trust it because I have submitted to the reality of who He is.

This is what getting out of His way looks like. I can't lean on my own understanding here (Proverbs 3:5-6). My understanding is based on *worldly* ideas of goodness, fairness, equity, and what I can see with my own eyes. My understanding of the world is innately tied to a need for justice and vindication. I want balance and tidy resolutions. My understanding finds a deep wrongness in the absence of these things. After the broad-daylight attack on my family, my human understanding tells me that safety is an illusion, that goodness is no match for evil, that trying to invest in the world is futile when what you love can be ripped from you in the blink of an eye. If I don't get out of the way of God, I will be crushed by my own flawed and finite understanding.

Stewarding Your Story

Reality without hope is despair, and I choose to have hope because I trust who God is. He is the Great Author. His authorship does not just reign over the bigger history of the universe; He is the author of my story too. The hard part is that He's given me the pen. I get to wield the pen that tells the story of me. There is a lot of power in that. So much, in fact, that it is easy to imagine that I am in charge of it all. If I am not careful, I will negotiate a sharing or complete takeover of control with God on every circumstance. I so easily block the feel of His nudgings and take the credit for the blessings around me. I can even take the credit for making lemonade out of life's lemons!

In 1839, when Edward Bulwer-Lytton said "the pen is mightier than the sword," it was not a theological sentiment.[1] We can apply it here, though, to represent how our human minds wrestle control away from God. We get to write our stories down, and we get to interpret and commentate on them as we go. We get to deliver our histories or testimonies as we see them and wrap them in narratives of our choosing. We have free will and get to make choices. One of those choices is what we pour into our perspective and how we narrate our experience to whomever will listen. We have the pen. It is not a stretch to say that we regularly convince ourselves that our pen is mightier than God's Word, which the Bible calls the sword of the Spirit (Ephesians 6:17). After all, I can conjure the power of the pen any time I want to. Through the use of my words, I can determine what is felt, said, and written, especially in the age of the internet.

It's true. God has given me so much power over my own story. More important than the telling of it, though, is the living of it. The way I live is the ultimate translation of how I am writing my story. My first job is to ask where I am attempting to claim authorship. When I claim authorship, I am getting in the way and attempting to thwart God's purpose. We claim authorship in all kinds of ways. There are big ways, of course, like blatantly denying God's sovereignty or ignoring the need for a relationship with Him through prayer. We can deny that God is active in our lives or succumb to the enemy's suggestions that God either does not care about us or takes pleasure in our suffering. We can paint our God out to be a great punisher, suggesting that our badness is the catalyst for His actions against us. We can take credit for God's miracles or elevate ourselves above others in God's kingdom of service. These are easily identifiable thefts of God's authorship.

I often find myself stumbling into much more nuanced thievery, however. When I cling fiercely to my own expectations, I am in His way. When I refuse to budge from my carefully made plans, I am asserting my authority over His. When I ruminate—allow worry to fester in my head and heart—I am taking over as shepherd and protector.

When I judge myself and others by worldly standards, I am willfully undermining the author of all goodness. I am in His way all the time.

> The way I live is the ultimate translation of how I am writing my story.

God hasn't just given us the pen; He's tasked us with its stewardship. We each have a unique responsibility to humbly acknowledge His sovereignty and relinquish the illusion that we get to be the authors of what unfolds in our lives and our world. Our obedience to this arrangement is not required for His story to be complete. I can tell you, though, that our participation delights Him (Psalm 37). Our stewardship of the pen requires finding the balance between acknowledging His authorship and accepting our responsibility to translate it as faithfully as is possible, given our fallenness. It is a discipline of evaluating where, and why, I need to relinquish control and get out of God's way.

In grief and suffering, sometimes our survival instincts take over and we feel a powerful need to control as much as we can. Our brains want to protect us from further harm by minimizing the unknowns. We might start to edit or limit our relationships. We might tighten the reins on our children or spouses. We might cut out risky adventure or become more rigid with our routines. We might become intolerant to change. We might stop investing in the unpredictable outside world and become more reclusive. This need in the immediate can be unavoidable and can even serve a helpful purpose. Sometimes we need to pull in or isolate in order to assess and accept the damage. We need stability during these times more than ever, and that is an okay place to be.

After my family tragedy, going to church was hard. It felt impossible, even, at times. I grew up in those pews next to my parents. I became a mother in those pews and got to watch my family grow. My church family not only knows my pain but has its own intimate grief associated with the shooting. It was hard not to sit in those pews and feel the presence of my lost family so acutely. It was hard not to hear their voices in the beloved hymns. It was impossible not to openly cry. It was definitely difficult to be willing to expose my grief in that way. And so often, I didn't. When my friend Julee was checking in on me and I shared this with her, I remember her response being so gracefully profound. "It is okay to be there," she told me, "but you can't stay there." I admit that even as I write this, I still experience difficulty sitting in the beloved church of my youth and adulthood because of the acute emotional connection to what I have lost, but it is something I am working on and through.

We have to be careful not to stay in retreat forever. There is absolutely a new normal that emerges as a result of grief, but we do not thrive in isolation or rigidity. As the shock of trauma, tragedy, and loss wears off, so does the fog of numbness. This can be an intense emergence. All the feelings are waiting for us. It is also a time when we can tentatively poke our heads into the now and start to venture back into the world we once knew so well. This is where the discipline of discerning between what is God's responsibility to hold and what is mine becomes more relevant. Where does God's authorship end and my penmanship begin? What can I let go of and give to God in order to find peace and meaning in my grief? What emotions, fears, or grievances can I lay at His feet? What stubborn blinders can I relinquish so I can actually see what God is placing in front of me?

In my own situation, letting go of the why has been a powerful wrestling characterized by submission. The why is part of God's redemptive story, not mine. What I think, feel, and write about the why, is that ours is a story of God's complete faithfulness. It is about mercy and love. It is a discipline to hold on to that understanding. In my weariness, I sometimes falter. Sometimes I sink into my humanity and wallow. There is plenty of room in that space for pity, resentment, and anger. There is no peace there, though; only bitterness. I think we can all agree that the search for peace is worth engaging, and that it takes work to make gains. Peace is a gift God has waiting for us if we ask, if we seek, and if we listen. Part of this searching is relational, and the voice we choose is important. The words we use for our own experiences matter a great deal in this endeavor. Yes, semantics matter. My prayer is that, God willing, from the tip of my pen will always come a story told through the lens of His victory.

Reflection Questions

1. How have you felt let down by the world? What perceptions did you have as a child about the world of grown-ups that you were disappointed to find out were not true?
2. What kind of relationship do you have with boundaries? What might you need to do to be better at understanding them and giving them integrity?
3. What part of your story might God be using to further His kingdom? What kind of story are you telling through your words and actions?

CHAPTER 6

Tackling the Unthinkable

Traversing the Waters of Forgiveness

At first people refuse to believe that a strange new thing can be done, then they begin to hope it can be done, then they see it can be done—then it is done and all the world wonders why it was not done centuries ago.

Frances Hodgson Burnett

At the time of this writing, a quick online search for available books on the topic of forgiveness will turn up more than 20,000 titles. That's a lot of words and ideas about such a ubiquitous interpersonal concept. Scripture tells us plainly that we must forgive as we have been forgiven (Colossians 3:12-13). One of my dad's favorite authors, Mark Twain, allegedly wrote that "forgiveness is the fragrance that the violet sheds on the heel that has crushed it."[1] Enlightenment-era poet Alexander Pope famously said, "to err is human; to forgive, divine."[2] The idea that forgiveness is a necessary part of relationships is a common human understanding. The phrase "say you're sorry" is ingrained into the cosmic parental lexicon so we can teach accountability and model forgiveness for our children. The phrase has become so common that it can fall into the category of rote, routine, and, in worst cases, meaningless. We engage in the practice, nonetheless. But is it that easy to forgive? Is an apology all it takes? Even with the extensive commentary we have, how many of us know how to truly accomplish this mandate?

Once again, this is an issue I repeatedly come up against in my practice. Survivors of emotional wounds, wounds that would often be fatal if they were physical, come to my office for help moving through what was done to them or what they have done to others. Whether or not they are Christians, they come in with an underlying understanding that forgiveness is part of that process, and they don't know how to do it. I humbly admit that I can't give you an exact blueprint of how to walk someone through the process of forgiveness, if there even is a process. It is deeply personal and complicated, but I do know how to clear away some of the debris that gets in the way of moving toward actual and lasting forgiveness.

Grace and Forgiveness

On the front end, differentiating between forgiveness and grace is important. Grace, as we are talking about here, foremost comes unmerited. Extending grace to others is to offer a gift or a blessing that is undeserved. When we are talking about grace as it partners with forgiveness, that blessing is often empathy or understanding. Another way to conceptualize grace is a withholding of warranted judgment in favor of unwarranted, unconditional love.

God's grace to us comes in the form of His ultimate and perfect sacrifice, Christ on the cross. This was a gift given out of complete and perfect love to a people who are utterly undeserving. Charles H. Spurgeon, in his 1889 sermon "Grace for Grace," said that "Grace does not begin halfway down the alphabet, but it is the Alpha of our hope."[3] It is at the core of who God is and how we are brought to His feet. Grace is a beautiful and unique gift that we first receive, in its perfection, from our Father. Hopefully, that grace then extends outward from us as love and compassion.

Forgiveness is a related but different concept. We can approach the whole person with grace, but forgiveness is specific to a certain behavior or action. Forgiveness is a mechanism designed to address transgression. Grace is for the person themselves, while forgiveness is attached to a thing we, or someone else, did. Forgiveness is also complicated. Because it is attached to a thing, we can justify withholding it based on someone's worthiness. We can allow forgiveness to hinge on the authenticity of someone's regret or the articulation of their wrongness. We are not necessarily dismissing their personhood by withholding forgiveness but are allowing it to be attached to a performative earning. Forgiveness can feel justifiably transactional next to the unmerited quality of grace.

We talk about forgiveness in a book on grief, loss, trauma, and struggle because unforgiveness is a clinging, living thing. It is an extension of the wrong that we have done or was done to us. It attaches to us and continues to inflict pain. Unforgiveness is profoundly personal. It buries itself deep within and calls the shots from those hidden depths. Because it has a say over our thoughts and behaviors, unforgiveness is connected to who we are. And who we are on the other side of this grief journey matters. Multiple parts of the journey through the shadow involve wrestling with self, others, meaning, and desire. Forgiveness is an inevitable part of the wrestling. Even if there is no one to blame, if your loved one passed peacefully in the night, or if a tornado struck suddenly and without warning, a wrestling takes place. In the absence of that wrestling, we accept unforgiveness as a way of being, and with it we accept the bitterness, resentment, and emptiness that comes with it.

The man who murdered members of my family was a lifelong football player whose career ended in the NFL. At the time of the murders, he was living right down the street from us with his parents, although we did not know them. I drove past his home almost daily. I must have seen and waved to him before. He must have seen my mom-van full of kids drive by. For how many years did we pass one another before the afternoon he killed my parents, niece, and nephew? How did I not feel it? Sense it? How was there no warning?

I was with my brother and sister-in-law right after we found out what happened to their children and to our parents. Those intimate, heartbreaking, and out-of-time moments are sacred. The first prayer I heard my brother utter after the assault was an anguished, loud, and guttural call for forgiveness for Phillip Adams. Forgiveness. Forgiveness for the man who killed his parents and children. I heard those words, allowed them to crescendo in my head, and I remember thinking, *Okay, this is how we are going to do this*. And so we did. And so we still are.

Forgiveness is a wrestling, and that wrestling is a choice. It is a decision about the way we are going to move forward. It is not a lightning bolt that wipes away what was done. It is not an erasure that makes the past disappear. Forgiveness is a gathering of thoughts and words that become the wrapping for transgressions committed against you. This thing that has happened to you does not go away, but forgiveness means that you decide to cradle it in something of your choosing so that it no longer pricks you with bitterness. Forgiveness is not an undoing, but rather a purposeful reconstructing.

Forgiveness is not an undoing, but rather a purposeful reconstructing.

Letting Go

The first hurdle to overcome when wrestling with forgiveness is letting go of the need for vindication. By nature, I am a fighter. In the raw, I am contrary, rebellious, and outspoken. I am also very justice oriented. When I first began working with trauma survivors, I kept finding myself getting caught up in my own anger and outrage over the lack of justice for victims. I would often storm into my mentor's office and vent about the unfairness of it all. My mentor, Bev, once told me that my job was to tend to the sheep and let God handle the wolves. I remind myself of this weekly, if not daily.

If grief is a journey, and forgiveness is part of that journey, this is the place where I reluctantly let go of something that feels so important: vindication. My soul needs for systems to do right, show up, and shine bright lights in the shadows to expose corruption. My soul longs for this because we were made to abide in the light. This world is fallen, though, and as much as I want to create a perfectly just and safe place, I cannot. I too am fallen and corrupt. My longing for justice is righteous, but my judgment is not reliable. Our peers, communities, and systems, by extension and regardless of intention, are equally as unreliable. Sometimes we get to have vindication, but often we are left wanting. Sometimes we get the apology, but often we are left holding the remnants of wrongdoing while the offenders walk away.

Phillip Adams shot himself that night. Our chance for a discourse that might give us some indication of why he did what he did died with him. Our chance to lay bare the depth of destruction that he inflicted died with him. By the route of communication with Phillip Adams, there is no opportunity for justice or vindication in my family's story. Certainly not in the earthly way that we might want. If I hold on to unforgiveness for something so depraved and heinous, I might die from the stress and weight of it.

To be clear, I am not suggesting we do not advocate for ourselves or that we abandon the battle for what is right. As I said, I am a fighter and have yet to be called out of the ring. Letting go of my emotional need for vindication, though, frees me up to think clearly and put my energy where it will be most effective. When my head and heart are dragging the corpse of unforgiveness around, my soul is tired, my vision is blurred, and I have trouble assessing what my next steps should be.

Clearing the Way

Another way to think about this process utilizes the language of debt. If forgiveness is a wiping away of debt, there are two steps I need to take to clear the way. First, I need to decide that I am going to allow God to collect this debt. I relinquish my earthly right to exact payment for what is owed. I choose to rely on the promise that God will hold my transgressors (and me) to account (Romans 14:12). This applies when we are working on forgiving others and ourselves. I need to allow God to collect when and how He sees fit.

This is not to say that we should not repent of or atone for our transgressions. This does not mean that we should not apologize or seek reparation for damage we have caused. Those are necessary disciplines. I am talking about allowing God to take the seemingly ubiquitous need for self-flagellation from us. Give Him the self-loathing, lay it as His feet, and trust that He knows what to do with it. God can hold and handle it. If we keep it for ourselves—covet it, even—it will stand between us and Christ, creating a barrier of shame. That falls under the category of spiritual warfare, and the stakes are high.

Submission

There is another, equally important step here. For me, the act of submission is more difficult. I need to carefully consider my own capacity to even calculate what is owed in the first place. Again, my judgment is not reliable. I only have two eyes, and they see first what my own experience or my own trauma dictates. If I am lucky, I make some space for context provided by others. Honestly, though, if I had the power to pronounce judgment on those who have hurt me, I wonder how many bodies would be in my wake. Don't misunderstand; I do not wish anyone dead. But I also don't trust that my emotions can be reliable when they are so big, so reactionary, and so fallible. I also cannot trust my perspective to account for all the nuances that contribute to people's thoughts and actions. I just don't know what they have been through. The only thing I know for certain is the measure of my pain and shame, and, at times, my pain has felt unsurvivable. If I can't trust myself to assess, shape, and declare the debt, then I must also give that to God. Not only do I give the debt to Him to collect, but I humbly acknowledge that I am incapable of measuring the debt in the first place. I also give that to the only One who can.

So I determine to allow God to have the final say over the wolves. This has stayed with me over the years and guided me in my vocational purpose. Also, I am a sheep. I wander. I get lost. I need tending. One of my jobs is allowing the great Shepherd to lead me. Another of my jobs is tending to myself in part by guarding my thoughts and being deeply protective about what I allow to simmer in my heart. Ephesians 4:31 tells me to "get rid of all bitterness, rage and anger, brawling and slander, along with every form of malice." Notice that it does not differentiate between feeling that way toward others or ourselves. The mandate is the same either way. I wish the snap of a finger could result in simply getting rid of all that negativity. The next verse tells us how. Scripture comes alive here to show itself as the solid basis for forgiveness psychology. The next verse, Ephesians 4:32, calls for compassion and forgiveness. Scripture connected bitterness, rage, and malice to unforgiveness before psychology ever did. In other words, forgiveness is the path to being free from bitterness.

I am valuable to God—a treasured possession, in fact. It is my job to be a good steward of the treasure that I am to Him. Part of that stewardship is guarding my head and being careful to cultivate what is written on "the tablet of [my] heart" (Proverbs 3:3). Forgiveness is about the work of curating what I allow to grow inside of me. This is the business of the innermost circle, where I exist with God. When distilled down to its core, forgiveness is about my relationship with God and the soul I tend on His behalf. It is worth it to me to forgive others. It is worth it to me to forgive myself. I am worth the hard work that is forgiving.

What You Sow in Your Heart

The object of my forgiveness is a more difficult matter to wrestle with. Again, we often associate forgiveness with the worthiness of the offender. Do they deserve my forgiveness? From a human perspective, it is often an insurmountable hurdle. Have you ever openly declared forgiveness only to find your heart still holding on to bitterness, resentment, or anger? I have too. Have you ever come face to face with the mandate to forgive others, chosen to obey, but then found yourself unable to follow through? Same here. Do you know what it's like to look at yourself in the mirror and want to forgive yourself, only to be accosted by an internal litany of reasons why you don't deserve it? Yes. A lot of times it just doesn't seem to stick.

This is the journey of forgiveness. If we make it about our oppressors' deservedness (or our own), we will never see the fruit. I believe God mandates us to forgive for several reasons. He wants us to understand the nature of His love for us (1 John 3:1) and His willingness to wrap us in Christ (Galatians 3:27) so we can be close to Him (Hebrews 10:22). He wants us to model for others what His love looks like (Ephesians 5:1-2) so they can recognize Him through us (Matthew 5:16). He wants us to experience His inexplicable peace through the gift of love (John 14:27). He is jealous of our hearts (Deuteronomy 4:24) and cares what we allow to grow in them (Proverbs 4:23). Maybe the following reason He mandates forgiveness is the simplest reason of all: *God knows it is good for us* (2 Corinthians 2:7-10; James 5:15-16; Mark 11:25-26).

It is good for me to let go of bitterness and malice. It is good for me to rid myself of the resentment that comes with the need for vindication. It is good for me to lay down the burden of the wolves and tend to my sheepish heart. When I replace those things with peace, compassion, and hope, it is good, and less stands between me and Jesus.

So we take those transgressions and we view them with eyes wide open. We do not bury them, sweep them under anything, or refuse to acknowledge them. Instead, we view them for what they are and radically accept the offense and the damage done. This thing that was done to me exists. The damage I caused is real. I will not deny the truth of the hurts that have been rendered at my expense. As my dad would say, "It is what it is." Now we get to decide what to do with it. This is where our power comes in. We get to choose where to go from here.

There is no denying that it helps tremendously when our offender comes to us with a contrite and repentant heart. It feels comforting when the people we have hurt offer forgiveness and reconciliation. It helps when we do get vindication in one form or another. Even if those who hurt us do not deserve our forgiveness, it helps the process if they at least acknowledge that we have been wronged and that they are sorry. Someone's contrition, however, is also not the secret sauce for our ability to truly forgive. I am not God. My humanity struggles with balance, finiteness, and the truth that I can see with my eyes. The human in me keeps ledgers. I need a reason to be disciplined enough to overcome my humanity here.

And what about the transgressions for which we will never get an apology? The hurts done to us by shadows from our past. Wounds inflicted by people who will seemingly never see the hurtfulness of their actions. There is a black hole somewhere full of lost "I'm sorrys." This is where it becomes helpful that forgiveness does not hinge on deservedness, contrition, or accountability. Let us unburden ourselves from seeking or proving those things. This is where we step away from other-centered forgiveness toward soul-centered forgiveness. The health of my soul feels like motivation enough for me to do the hard work of wrestling with self.

The salvation of my sister's soul is not bound to my forgiveness of her. It is bound to God's. The salvation of my soul is also not bound to my forgiveness of my sister, but it is affected by it. The quality of what I sow in my heart and mind determines my thoughts, words, and actions. It is a soul issue. When I hover above my own story, and my decision to wrap what was done to my family in the cloak of forgiveness, it seems like a life-or-death choice. I can muster up an endless amount of anger and darkness in response to what has happened to us. The quality of my life moving forward, though, rests on whether I steep my soul in love or in anger.

> The quality of what I sow in my heart and mind determines my thoughts, words, and actions.

Forgiving Phillip Adams is a choice between life or death, and I have had enough death. We have all had enough bitterness, anger, and resentment. We are all tired of disappointment. I choose to live. In that choosing, I unload the burden of vindication. I drop the requirement for contrition from anyone else. I rid myself of the weight of judgment. I don't want to hold the gavel. I can empty my hands of those things so that I can stand in that inner circle with my Savior and tend to what is really important. Now forgiveness has a chance.

Reflection Questions

1. What are some wounds that you want to forgive but can't? What areas of unforgiveness have stubbornly rooted themselves in your life?
2. What are you holding on to or holding out for that you think would help you forgive? Is there a way you can move forward in forgiveness without those things?
3. Can you commit to making forgiveness part of your lifestyle? What might stand in your way of staying true to this commitment?

CHAPTER 7

Regret

Facing Hard Things, Owning Them, and Letting Go

Where have you buried your best days? Have you lived or not?

Fyodor Dostoevsky

Our theoretically grown-up brains have the capacity to negotiate perspective on a litany of daily issues. This mental maneuvering is often lightning fast. In the space between stimulus and response, our brains apply context and common sense to determine how we might think about or react to a situation. If we are disciplined, we apply reason, rationale, and values to our perspective. If we are thoughtful, we can sift through the facets of truth and reality to settle on a perspective that is, at the least, not destructive. This rapid process allows us to make the kind of quick decisions that are required for executive and higher-level functioning. It is the kind of split-second analysis that helps us determine how to react in any circumstance. This process helps me get out of bed when my alarm goes off or respond to physical sensations my body is signaling like pain, hunger, or cold.

It also functions in a slower, more in-depth way that we can think of as critical thinking. This complex thought process allows for humility, accountability, grace, and forgiveness. Conversely, the result can also fertilize seeds of mistrust, contempt, bitterness, and resentment. Consider the spilled milk example. Do I cry over it? Does it ruin my day? Can I take it in stride? My husband leaves his socks in the strangest places. I think I've found one in the pantry before. Is this habit a marriage-ending annoyance? Is it adorable? In the moment, my brain pulls from a myriad of resources to help me decide how to react. Do I storm down to his office with laundry-leaving evidence in my fist, or do I lovingly leave the sock as a secret treasure for him to discover and collect later?

When something traumatic happens, our typical, well-honed systems of response are often compromised. Sometimes, in the heat of the difficult moment, our most basic automatic responses take over without the more complex assistance of critical thinking. Our default settings are often rudimentary and childlike. These might be what we consider knee-jerk reactions. Essentially, our initial responses to emotional discomfort or pain carry with them all the lessons from our past: good or bad, functional or dysfunctional, constructive or irrational. Our responses represent the culmination of our life experiences and worldviews. This represents a lifetime of lessons, interactions, thoughts, takeaways, impressions, injuries, or triumphs. When the dust begins to settle, it can be difficult to make sense of how these different and sometimes contradictory pieces have influenced our reactions.

Perspective Negotiating

Perspective negotiating, or processing, becomes a necessary analytical endeavor when we need to figure out why our brains did what they did. Where did those thoughts, behaviors, and automatic responses come from? I once had a client refer to me as the "great egg-unscrambler." I don't know if I can claim that title, but I know what he meant. The exercise of untangling your internal processes is hard work. Childlike voices rage to be heard in our discomfort and pain. Neglected parts of us sense vulnerability in our adult armor and try to break free. Our hurt parts come out to radiate their insecurities. These are the parts of us that jockey for the driver's seat when we try to make sense of pain. Perspective negotiating requires being aware of the influence or neglect of these voices and discerning their appropriateness—egg-unscrambling.

Shame

One of those voices is almost always associated with shame. I do a lot of work with clients around shame. It is unavoidable in my field. It is unavoidable in our heads, for that matter. Shame first makes its appearance in human history way back in Genesis 1. Shame was the first, and predominant, emotion experienced by Adam and Eve immediately after disobeying God and eating the forbidden fruit. Shame was instant, and their instinctual response to feeling it was instant: They hid. They hid their nakedness (or vulnerability), they physically hid from God Himself, and they hid the truth. Shame is that way. It encourages us to hide. *Shame entered the world with sin, and we have never been able to shake it.*

There is also nothing redeemable about shame. It undermines our worth, it lies to us, and it dehumanizes us. It marks us with disgust and disgrace. It tells us to hide from the people who love us the most. It convinces us that the truth makes us vile, so the truth must be dangerous. While guilt tells us that we *did* something wrong, shame tells us that we *are* something wrong. It operates in shadows and deception and requires darkness to flourish and function. Shame is our enemy, but we often hold it close in a death grip.

I do want to argue semantics here and differentiate between internalizing shame and feeling ashamed. While the root word is the same, I would argue that feeling ashamed is a deeper extension of guilt and can have a constructive purpose. Feeling ashamed of our behavior can and should draw us into the light of accountability and true repentance. Internalizing shame, however, is like a bewitching voice calling us into a dark alley, ushering us into hiding. Hiding leads to denial, isolation, and an extension of dysfunctional protective tendencies.

> While guilt tells us that we *did* something wrong, shame tells us that we *are* something wrong.

Regret

Now let's talk about regret. Like guilt, regret has a functional purpose. It is a negative feeling, for sure, but can serve as a corrective reminder of poor choices. *If guilt emerges as an alarm that we have missed the mark, regret lingers to inform and shape our future choices.* Regret can be an appropriate admonitive presence that accompanies remorse and accountability. If the ultimate result of regret is insight, repentance, and positive change, it has served its purpose. When considering the positive aspects of regret, consider scriptural mandates to remember your story and continue to apply its lessons to your life (Deuteronomy 4:9, 8:2; Psalm 143:5; Isaiah 46:9). We are not to forget or undermine the lessons we have learned, especially the insights that have been forged through hardship. We are to remember them with clarity so we do not repeat our mistakes.

What happens, though, when shame meets regret? What happens when these emotions collide and invade our internal processes? As constructive as guilt and regret can be for shaping character, integrity, and future behaviors, shame and regret can be equally destructive to them. If shame emerges to undermine our self-identity, regret latches on as a virus that corrupts confidence in our ability to act. Regret stops being functional and becomes an albatross. Instead of using our past mistakes to free us from future mistakes, regret, when it is tied to shame, shackles us to the darkness and the behaviors that thrive therein.

Every once in a while, a specific memory from seventh grade pops up in my head. It is so irritating. I was probably 13 years old. I was standing outside at lunch with a group of people. I said something dumb. It wasn't rude or offensive, but it absolutely showcased my cluelessness. A couple of the guys I was talking to gave me a side eye, and then the conversation simply moved on. I am sure no one remembers this but me. (If you were in that group, are reading this now, and remember that moment, I respectfully ask you not to tell me.) And yet, decades later, when I least expect it, I cringe. That's not the only memory I have that has embedded itself in my regret bank. And that regret does nothing for me. It makes me feel stupid. When it pops up in my head, that feeling of embarrassment is right on its heels. There is shame there, and it won't let me forget. Those memories spin me around in circles. They are not constructive but rather act to destroy my sense of self-worth in the here and now.

A lot of those kinds of embarrassing memories from our developmental years are benign. There are some that are worse, some that cut deeper into our ability to love ourselves. Some of these shame-encrusted memories can cover us completely with self-loathing when they emerge. We can all identify with the heavy and sometimes dirty feeling of ruminative regret. These regrets embed themselves in our psyches and settle in for the long haul. These are the regrets we have to put energy into working through in order to dislodge them. These are the ones that require a severance between shame and regret. They take active work to sort through.

We have to exercise all our perspective-negotiating skills to replace shame with guilt, sort through the context of the situation, and allow ourselves to have the peace that comes with self-forgiveness. This is the goal here: forgiveness of self and others so we can move forward with the kind of truth and grace that act as formative guides.

I might be biased, but therapy can help with this process. Sorting through the complicated overlap of guilt, regret, and shame is often best done over conversation with a respectful, safe, and empathic ear. It is also helpful to sort through these things with someone who can tell you the truth with love. Shame hates the light of revelation, and putting language to our experiences is like holding them up to the sun. A safe and solid friend, mentor, family member, or therapist can be a great (and sometimes essential) place to start the process. Iron sharpens iron. Limiting the time we spend hiding in the darkness of shame is a benefit of good counsel and is worth the price of admission.

Repentance

Part of the recipe that allows us to take the benefit of regret, functioning toward insight and positive change, and then let the regret go, is our capacity to repent. Sometimes a simple apology will do, even if it's years after the fact. Repentance does mean committing to stop repeating the offensive behavior. Other times, reparations are necessary. We are talking about accountability here. A lot can and should be said about accountability. For our purposes, it is enough to insert accountability into the process of falling short, feeling guilt or shame, experiencing regret, and then finding peace. It is not enough that we know in our hearts we have done wrong. We need to do something with that knowledge. The instrument that moves the needle from knowing to doing is accountability. Accountability is more than an idea; it is an action, a commitment, a humbling. Accountability requires a reconciliation.

When we apply accountability to regret, we have the power to genuinely move forward. Determine what responsibilities are yours to account for and find your courage. While you are at it, let go of the responsibilities that belong to others. There are two sides to this coin: Hold what is yours to hold and stop carrying what belongs to those around you. Accountability is hard enough when we aren't accepting blame for things that aren't our fault.

> Accountability is more than an idea; it is an action, a commitment, a humbling.

Holding the mistakes or faults of those around us is a dangerous practice. First, we need to recognize that allowing others to take the blame is a common human behavior. We hone that practice early and effectively. The problem is we don't always stop this behavior when we get older. The professionals call this gaslighting and consider it emotionally abusive. When I blame you for my mistake, make you think it is your fault, and then hold you to account for it, that is gaslighting. Sometimes we take on the mistakes of others willingly. In my practice, I often work with adults trying to give accountability back to their parents for childhood traumas. We hold a lot, especially as children, and the work it takes unburdening ourselves from the responsibility of other peoples' dysfunction is worth the sweat.

Our energy is precious, and our willingness to admit mistakes can be precarious. Part of protecting our ability to face our mistakes is guarding our resource pool. We don't have the bandwidth to carry our mistakes on top of everyone else's. My dad always said that we only have so much emotional coinage to spend. He is right. We need to be discerning about what we choose to carry so, whatever we choose, we can carry it well and all the way home.

This is the process of dealing with regret. We first acknowledge that the source of our discomfort is shame or guilt. If it is shame, we should confront the impulse to hide, shine a light on our experience, and allow it to transform into its more functional counterpart, guilt. We then determine what, in this circumstance, is ours to hold. Ask yourself: *What offenses are mine to account for? What parts of the context at hand belong to others or are things outside my control?* Hold what is yours and let go of what isn't. Allow for conviction to give way to accountability and follow through with what that entails: apologize, atone, repent, confront, and forgive. Simple, right?

Complicated Grief

Yes and no. If regret were simple, I wouldn't be writing about it. In the best scenarios, regret requires attention and work. What happens when it gets far more complicated? What happens when our opportunity to make amends or change behaviors is suddenly ripped from us?

Psychologists describe *complicated grief* as prolonged and intense grief that gets in the way of healing. Some of you might be thinking that this description sounds like normal grief, and I would agree. The death of a loved one is often world-shattering. Losing dreams, trust, or our security is a crisis. And that is just the initial impact. The subsequent months and years can feel like nuclear winter. On top of the grief is sometimes the confusion of complex relationships, disconnection, unresolved issues, estrangement, or simply things left unsaid. When these get in the way of healing, when regret becomes an albatross, this can lead to complicated grief.

God is all about the "re" words: repentance, reconciliation, renewal, redemption, resurrection, restoration. These words all have one thing in common: *re*lationship. I believe that the inner pull toward repairing our relationships is divinely inherent. Throughout Scripture, we are shown paths toward God and toward one another. Isolation and alienation are described as forms of earthly punishment (as with John on Patmos), divine retribution (as with the exile of the Israelites), or periods of concentrated wrestling with something significant (as with Jonah in the belly of the whale or Jesus in the desert). The ultimate exile is separation from God. We are always meant to find our way back, to make it to the promised land. We are prodigal sons. Reconciliation has been the plan since the fall of man.

In this light, it is understandable that estrangements from loved ones are especially painful and unsettling. They are painful whether that separation is for the best or not. It is not the way it is supposed to be. That discomfort can turn into torment when we lose the possibility of reconciliation completely through death or any other insurmountable complication. You are not alone if you thought you would have time to fix it. So what do we do when our grief is complicated by regret?

To be clear, this kind of regret does not necessarily coincide with our own transgressions. We can regret wrongs that were done to us, or things left undone completely. The same concept applies either way. I might regret missing my chance to confront a loved one about the way they hurt me. I might regret not telling someone that they are forgiven while they were able to hear it. No matter who is at fault, we are left holding the unfinished. It's not fair. That is a good place to start: acknowledging the weight of the undone, the heaviness of that unfairness. The impact of that alone can draw out a myriad of emotions like anger, resentment, frustration, helplessness, loneliness, or fear. Reject the impulse to run from these feelings or numb them and just sit with them for a minute. Put language to those feelings. Give them a place to go. Put words to them so you can understand them. If we do not do this work, the destructive nature of unforgiveness has its chance to wreak havoc. Take that unfairness to God and wrestle with it in His presence. He can take it.

Emotions are informative, not directive. Just because you feel something does not mean you have to or get to act on it. But your feelings are very important. They give you vital information about what's happening around and in you. Respect them by listening to them and allowing them to take shape. This is a good time to enlist an empathic ear. Bring in your friend, family member, mentor, pastor, or therapist. Give your words a safe place to land. No one needs to talk you out of your feelings—these are your emotions, they are valid, and they are giving you good information.

Emotions are informative, not directive. Just because you feel something does not mean you have to or get to act on it.

What to do with that information is the next step. Okay, so I am angry. This is unfair. It hurts. Wrong things hurt. How do I make a wrong thing right when the other half of the equation is out of my reach? How can I repair this broken situation all by myself? This kind of resolution requires higher-level soul searching and perspective negotiating. We need to once again ask and answer the question, "What is mine to hold here?" What wrongs have I committed and what rights have I omitted? Answering these questions gives us actionable steps to simplify the situation. As the famous proverb of unknown origin asks, "How do you eat an elephant? One bite at a time."

Why is also important. What context surrounds my decision-making, and can I make sense of why I made the choices I made? We rarely withhold important words or actions just because we are stubborn or ignorant. Usually a lifetime of hurts, happenings, and habits come into play when we choose a path. A lot of these influences are engaging with our subconscious and not up front where we can see their contributions. Remember the subconscious processes from the beginning of the chapter. So much is at play under the surface.

This may involve some egg-unscrambling. If we no longer have the option of confronting the object of our disconnect and hashing things out, we need to figure out how to fit the pieces together ourselves. Again, this requires a willingness to do the work. It is not always comfortable to face our past and what has contributed to who we are now. Squaring up to self is rarely easy. We do this work, however, to unburden ourselves from the heaviness of unresolved disconnect or unreconciled trauma. These burdens are the antithesis of peace and can certainly inhibit personal growth.

My relationship with my mother was sometimes complicated. I was made of a lot of hard corners and rough edges as a young adult, and my mom wasn't sure how to relate to me. That created a distance between us that waxed and waned throughout the rest of her life. Blessedly, we were able to have hard and important conversations in her later years and spent the last several years of her life in a positive relational space. We figured out how to understand and express love to one another. It is a cherished gift.

I also have had to wrestle with my own regret in our relationship. I thought I would have forever to completely figure things out with her. Now that she's gone, I see missed opportunities in so many of my memories with her. In fact, my longing for treasured and redeemed time with her makes the reality of our relationship a bit hazy. If I am not careful, I can put the burden of all that disconnect on myself. The truth is, she wasn't perfect. She wasn't a perfect mother. As I tell my clients all the time, though, if our parents were perfect, we wouldn't need Jesus!

I am still looking at myself in the mirror and trying to be honest. I am still reckoning with my shortcomings and the tragic implications of their impact on my relationship with my mother. It is a matter of integrity for me to have the courage to face and account for how I have hurt her and others. It is up to me to take what I find there and propel positive change into my present and future relationships. That is my job.

Doing my job is how I unhook myself from the shackles of shame. I choose to allow regret to be my teacher, not my tormentor. The work can be complicated, but the reward is positive growth and change. Let the great egg-unscrambling begin!

Reflection Questions

1. Are you open to "perspective negotiating" some areas of hurt or regret in your life? What might be hindering you from opening yourself up to a more nuanced perspective?
2. When you think about what areas of shame you are holding on to, what steps do you need to take to shine a light on them to alleviate the power of destruction they have over you?
3. What specific hurts have you caused that might benefit from explicit accountability? What is preventing you from taking steps in that direction?

CHAPTER 8

Loving Your Neighbor

A Closer Look at Our Jobs as Caregivers and Grievers

"I was hungry and you gave me something to eat, I was thirsty and you gave me something to drink, I was a stranger and you invited me in, I needed clothes and you clothed me, I was sick and you looked after me, I was in prison and you came to visit me."

Matthew 25:35-36

Sometimes I come across a social media post titled something like "What Not to Say to Someone Who Is Grieving." Always on the list are statements like "they are in a better place" and "it is going to be okay." These lists typically make me cringe. How many times have I accidentally stepped in it by saying the wrong thing myself? I get why the lists are written, and I even understand why some of those words might fall short. When I glance over those admonitions, I feel discouraged. I am certain I am guilty of saying unhelpful things to people in pain! The simple truth is that we often have no idea what to say or do when tending to someone who is deeply suffering. Figuring out when and how to step in can be a tricky endeavor.

Caregiving can be simple and beautiful, like a casserole or a hug. It can be uncomplicated, like helping with carpool or mailing a thoughtful card. Caregiving can be practical. For example, my mom used to sit with one of her dear church friends during chemo treatments. This was kind, pragmatic, and helpful. Caregiving can also be messy and difficult to navigate. If this wasn't the case, those lists of what not to do wouldn't exist. Caregiving gets less straightforward when the person in need isn't in our inner circles, when we are unsure how to tend to those who are closest to us, when we don't understand the grief, when the level of loss is scary or unknown to us, or when the timer goes off and we determine that the grief or suffering period should be over.

Who Is My Neighbor?

Let's start with the first complication. What is our job when the person in grief isn't my sister or my best friend? What if I barely know them? Is it even my job to tend to those people in the first place? Would they even notice if I didn't show up? Would they even want me near their pain? Who, exactly, is my neighbor? The parable of the good Samaritan (Luke 10:25-37) is one of my favorites. It has danger, intrigue, surprise twists, and a pretty happy ending. Let's recap. In this parable, Jesus tells the story of an implicitly Jewish man who is beaten, robbed, and left for dead on the side of the road by robbers. (I picture him in a ditch, but I don't think that's in the story.) Multiple people pass by (fellow Jews who are supposedly his neighbors) and do not help him. The passersby weren't just lay people. One was a priest, and one was a Levite. In other words, these were people who should be theoretically inclined to help a fellow Jewish man. The story doesn't end there. I can almost hear the suspenseful music start up when a Samaritan then enters the story.

A little context goes a long way here.The Jewish people did not like Samaritans. As a matter of fact, there was such long-term beef between them that the Jewish people would walk around Samaria just to avoid them. Political, religious, and ethnic differences caused deep-seated mistrust and animosity between the groups. They weren't simply unfriendly; they were enemies. This context adds such depth to the parable. Back to the story. The music is getting more tense as a Samaritan approaches the helpless, hurt, and dying Jewish man. What does he do? Spit on his enemy? Kick him? Laugh? Does he determine that this is definitely not his business? Get this: He loves him. He assumes the role of caregiver. He stops, cares for him, takes him to an inn, and arranges for ongoing help.

Jesus tells this parable in response to a question regarding eternal life. Jesus says that to inherit eternal life, you must "Love the Lord your God with all your heart and with all your soul and with all your strength and with all your mind" and "Love your neighbor as yourself" (Luke 10:27). The story doesn't happily end at the inn; it begins and ends with a mission from Christ: Go and love your neighbor as yourself.

There are two parts to that command. The first part is to love your neighbor, but let's start with the second half of that mandate: love them *as you love yourself.* Let's consider how we love ourselves. Self-worth is something else I deal with a lot in my practice. Do you love yourself? What does that love look like? Is it healthy? Active? Attentive? Gentle? Kind? Is that love generous and forgiving? Is it honest? Is there room for grace in that love? Or is that love abusive? Neglectful? Cold? Relentlessly negative? Unforgiving? Christ is clear in His mandate. Love is part of the equation! The way we love ourselves not only matters but also is a crucial component of our ability to love others.

This idea deserves some reflection. If the way I love myself is neglectful or abusive, does the love I extend to my neighbor also carry that stain? Is there an undercurrent to the way I love myself in every expression of the way I extend love to others? It is an interesting question. Is it possible to love our people well when we withhold that kind of love from ourselves? As I wrote about in chapter 4, this is inner circle work. Proverbs 4:23 says, "Above all else, guard your heart, for everything you do flows from it." The way I conceptualize my worth is a heart issue. If all I do, if the way I treat and care for others, flows out of my heart, it is important that I tend to what my heart contains. It is important that I guard my heart not only from unwanted outside influence, but also from the spiritual warfare that wages within. If I am loved so thoroughly that my sins were paid for on the cross, I must be very valuable to my God. It is a desire of my heart to love what God loves. That means me.

The way we love ourselves not only matters but also is a crucial component of our ability to love others.

I often tell my clients to stop and think before they speak to themselves. If you would not say something cruel to your friend, why on earth would you say that to yourself? If you wouldn't dare treat someone else in an ugly or unfair way, why are you doing it to you? Stop it. The way we love ourselves matters. We are all probably guilty of extending grace and patience with others while withholding those things from ourselves. We allow others to be messy and human but endlessly punish ourselves for the same struggles. At some point we adopt the idea that we don't deserve that kind of compassionate love. It is the ever-present whispered lie that floats across the spiritual battlefield. *You don't deserve love.* And it works. If we are convinced that we do not deserve to be loved, our capacity to love others diminishes. See the win there for the other team?

I expect the idea that the way we love ourselves impacts the way we love others to hit some of us hard. Many of us consider loving others our forte, even if we withhold that same love and consideration from ourselves. Loving others well is connected to the way I love myself. The implications of that connection are heavy. It is clear, though. According to Christ, we are to love our neighbors as we love ourselves—well. A lot more could be said about the journey of loving yourself, but I will simply say that it is worth exploring, considering, and analyzing. If you need help in that area, reach out to your people. In the meantime, start measuring your treatment of yourself against your treatment of others. If you wouldn't do it to your friend, don't do it to yourself. On the flip side, if you would offer something worthwhile to a friend—such as a kind word, compassion, or permission to rest—offer it to yourself.

Loving Your Neighbor

The more familiar part of Christ's command is to love your neighbor. The parable makes it clear that my neighbor is everyone. Even people I don't know. Even people I don't like. Even people who don't like me! Oof. There it is. My neighbor, who I am to love well, is all of you. It is our job to love those who are in our orbit. All of them. *How* we love our neighbors in grief does, however, depend in part on where in our orbit they fall.

My friend Joan is a victim's advocate in her soul and in her profession. She has a lot of experience navigating the various systems involved with violent crime, such as law enforcement and victim services. She also happens to be in my inner circle of friends. I knew nothing in those first few minutes, hours, days, and weeks after I learned what had happened to my family. I had no idea what was happening around me or what was coming. There was no way for me to understand what questions to ask and to whom I should be talking. I knew someone who could know those things for me, though: Joan. In the blur of sirens, badges, and the wailing of my loved ones, I called the only person I knew who could love me in this very specific way. And she did. For Joan, loving me meant using her gifts, experience, and connections to advocate for my family when we needed something solid and familiar. She knew her capabilities, honored them, and used them to love us. She was able to show up for us in a definitive way that no one else could.

My friend Hannah left her husband and children, got in a car, and drove to a hotel near my home just so she could be within reach. I remember her text saying she was close if I needed her, which I did. She stayed for a week. At the time, I was too overwhelmed to realize the sacrifice she was making to be with me during that first week. Her place by my side seemed unquestionable. From her heart came the knowledge that her presence might have value to me, and she made that value easily accessible. She was there for me to grab onto when I needed a solid rock in the storm.

My house became ground zero for all things grief. My door was never closed, and my house was never empty. My friends Nora and Ginger showed up over and over, quietly, with food, coordinating the contributions of so many others who wanted to help but didn't know how. Our pastors came to sit with and pray over us. Extended family (shout out to my husband's New Jersey family who rallied and gathered en masse) flew in, reached out, and made their presence known.

Others in my inner circle showed up just to sit with me, just to be near. I have friends who listened to my story over and over because I needed to hear the words come out of my mouth. There were people who just made me laugh, something I desperately needed. We received countless cards of support, many from total strangers. There have been tributes and memorials. Food and words. Love and care. Big things and little. The most important component in all of those things is intention.

Loving your neighbor in grief means simply showing up. It means being intentional. Again, how we show up is determined by our gifts and our relationship to the grieving. The lady down the street who has never approved of my driving does not want me showing up in her living room with an open shoulder to cry on. But she might appreciate flowers, a card, or a meal. It is easy to leave it up to those closer to the source to tend to the grieving. We can easily rationalize not getting involved. By *we* I mean *me*. I am guilty of watching from the sidelines and being okay with that choice. Loving your neighbor isn't a suggestion, though. It is a mandate. So those I do not know, know well, or even like get to be the recipient of my support even if they never know it. Prayer is the least and the greatest we can do for those people we might not be able to reach. Supporting from an appropriate distance counts and is felt. I promise.

Again, showing up is what matters. When we are closer to the grieving or suffering person, there are some things to remember. First, when it comes to what not to say, there is only one thing I would absolutely advise against: nothing. The worst thing you can say is nothing. Grieving people need to hear your voice and need to know you are there. Suffering is scary when you are in the eye of the storm. It is scarier to feel that your experience is so uncomfortable, foreign, or intimidating that people don't want to enter in. It is lonely in there.

Second, don't stop. My sister-in-law, Katie, has said that one of the most important gifts she's gotten since the death of her children has been the consistency of her people. They haven't stopped. Even when Katie takes space for herself, her people are nearby, and she knows it. Note that the care and help suffering people need should and does evolve. We may begin in crisis, but we don't stay in crisis. Our efforts to show up for others should change based on their need and what is appropriate. While consistency is key, the measure of how we are showing up might change. For example, while I appreciate every meal that was delivered to our family, there was a natural and appropriate cessation to that provision.

The third thing to consider is your fear. Fear keeps us from showing up. We fear saying the wrong thing or making the pain worse. People often tell me that they don't know what to say. My response to that is to say something anyway. You aren't going to make it worse. You aren't going to make it hurt more. What you are going to do is show me that I matter to you, show me that being my friend means more to me than your fear. That means a lot. Don't be afraid to use the words. Don't be afraid to say their names. Don't be afraid to enter the language of suffering. Grief can be lonely, and people don't want to be alone.

Another stumbling block we encounter when considering our neighbors is when we do not understand their suffering or grief. Some examples might be when someone's pet dies, when someone we know experiences a miscarriage or the loss of a dream. What about divorce, loss of a job, or relational betrayal? What about abuse? If we have been in those places, the suffering likely makes more sense to us. When the situation is foreign to us, we might not even consider it worthy of grieving. The answer to your confusion is to trust the experience of others. Your understanding is not required for someone else's experience to be valid. Show up anyway. Responding to pain is what matters, not your ability to sympathize or empathize. The good Samaritan did not question the wounded man before he reached down to help.

Your understanding is not required for someone else's experience to be valid.

Grappling with Grief

Grief, death especially, is uncomfortable. Our natural inclination is to put distance between that and us. We do not want to be near it if we can help it. One very hurtful narrative regarding the murders of my family was that Phillip Adams had been a patient of my dad's and that my dad withheld medication from him. People used this narrative to justify the murders. Not only is that claim unsubstantiated, but every aspect of it is false. People grabbed on to it, though. The comments on articles regarding this narrative were so hurtful with some even insinuating that my dad deserved what happened to him. I understand, though, why people believed it. It provided an answer in the absence of one. It was an explanation that made people feel safe. If I don't become a doctor and withhold medication from my patients, then I won't get murdered. If I can understand what happened, I can prevent it from happening to me. I get it. Without that false explanation, what happened to my family could happen to anyone, and that is terrifying.

The truth about our vulnerability is unsettling. We don't want to be reminded that simply being good and doing the right thing is not enough to protect ourselves or our loved ones. Our story is scary. That simple fact has kept people away and been the catalyst for people to simply "nope out" (as my sister Lori has said). Some people have just decided that they don't want to be near our story, don't want it to be part of their own stories. So they disappear from our lives, or nope out. I understand it. My story goes where I go. I can't get away from it, but you can if you want. It's tempting. Tempting, but not loving.

The correct response to the fear of someone's tragedy or grief is courage. Even though your inner voice might try to talk you out of it, you do belong down in the trench beside the suffering. It doesn't matter if you have experienced similar pain. It is okay if you do not understand. You are the Samaritan. Face the discomfort of being near pain and show love.

One more hurdle we face when caring for those in suffering is the dreaded stopwatch. We tend to allow just so much space for pain or grief and when it's used up, it's used up. This one really gets me. There are a lot of online opinions about the appropriate timeline for grief. The implication is that if you need an extension, you are not doing your job well enough. Maybe you aren't strong enough. Maybe you are stuck. Maybe you even like your suffering too much to let it go. Yikes!

Responding to the notion of being stuck in grief, my friend Joan says, "People processing grief are flowing back and forth like the tide…it isn't the same water each time. It might be the same shell popping up every once in a while—but it's not the same wave. It's literally changing!" Yes! Be careful when determining that someone else is stuck in their process. It might look like the same cycle to you, but there is often so much change and growth happening beneath the surface. It is also okay for it to be slow. As a clinician, I get nervous when people rush through their suffering. I understand it though. Who wants to hurt? But a hurried recovery can lead to so many more problems—emotional, behavioral, and physical—down the road.

The answer to so many things is grace. Acknowledge the limitations of your own understanding and proceed with humility. Give yourself permission not to know all the answers while you are giving the grieving person permission to simply be where they are. How to love them will flow from there, and it will be good.

Permission to be where you are doesn't just need to come from those who love and support us; it needs to come from ourselves. This is one of the jobs of the griever. We need to extend that grace to ourselves and just accept that our journey through suffering and pain is unique.

It is what it is. That acceptance can be tough. We put just as much, if not more, expectation on ourselves than we do on others. Comparing our journey to the journey of those in similar situations can lead to discouragement, but damage is a more likely outcome.

I am guilty of comparison as well. *If my sister is not having a hard time with this holiday, why should I be struggling through it? Other people have experienced worse. Why can't I seem to get over it? Maybe I shouldn't talk about my own struggle with someone who I know has lost more.* These are dangerous thoughts. We put ourselves at risk of minimizing our pain and experience, and we inadvertently isolate the object of our comparison in their pain and experience. This happens frequently. I will often, outside of work, find myself listening to someone try to express the shape of their pain, only for them to stop short and say something like, "I am sorry. I know that's nothing compared to what you've been through." Trust me, I'm screaming in my head: *No! Your pain is so valid! Please don't shut me out!*

When we hold ourselves to unreasonable standards, we cover the pain of grief with the agony of judgment. Addressing how we are not measuring up becomes the primary job, as opposed to working through our grief. It is like wrapping our injured and vulnerable body in a hair shirt. Pain on top of pain.

When we hold ourselves to unreasonable standards, we cover the pain of grief with the agony of judgment.

As grievers, our job is to simplify, and our internal systems long for it. Sometimes in grief we feel ourselves pull inward or experience the desire to make our circles smaller. Trust that intuition. This does not mean isolate, but it does mean allowing yourself to consider where you put your energy. We only have so much energy to give, and when our systems are in a constant state of internal wrestling, trying to make sense of the unthinkable, the extra energy we have for the world at large diminishes. It is okay to pull back a bit for a time. That means it is okay to say no. When someone asks you to head up the fund raiser, it is okay to decline this time. When your group of friends wants you to join them for dinner, it is okay to tell them you are not up for it. When it feels like the world needs you to show up and be okay, it is okay to say no.

A grieving client once told me that she sometimes wishes we still held on to some of the mourning practices of the Victorian era. Specifically, she said she wished she could be shrouded in black so people would see her from afar and understand that she is in pain, that she is vulnerable, and that she is doing the impossible. This makes so much sense to me. It is hard to be out in the world looking normal when my insides are on fire from suffering. Those around us want us to be normal. They want us to be okay. If we don't explicitly tell them otherwise, they will likely assume that we are. It is okay to say no, okay to admit that we don't have the capacity to tackle something, and okay to move through the world gingerly. Grief is delicate.

It is also okay and vital to say yes! Why is this just as hard for us? Friends, when someone asks you if they can help, say yes! A lot of us pride ourselves on independence and our ability to take care of ourselves and our people. It can be very hard to ask for or simply accept help. It might be an easier pill to swallow if we replace the word *help* with *love*. Can I love you? Is there anything I can do to show you love right now? Yes, please!

Semantics are important. Grief is often an unreachable or unhelpable state. So when people ask if they can help, the answer "no" comes from a deep understanding of our situation and emotions. Can anyone really help make this better? Help might not be able to take the pain away, but can love add something to the mix that becomes a hopeful part of our grief story? Multiple things can be true at the same time. Coexisting with pain, there can be joy. Mingling with a sense of loss can be the revelation of community. If this part of my journey is inescapable suffering, how sweet it is to be touched at the same time by care, comfort, and love.

Say yes to the offer of love. A casserole is not going to repair the damage of loss, but the love it represents can be a necessary reminder that our stories do not begin and end with grief. Cards might never be able to offer words that fully relieve us of our pain, but the thoughts and actions behind sending them tell us that we matter. Your people might not know exactly what you need when you need it, but their presence and willingness to show up can be the life preserver that offers us a chance to breathe.

Talk to your people. Say the words. Don't get lost when the world moves forward and you feel left behind. If you tell your people where you are, they have a better chance of being able to find you. Give yourself permission to be honest and give your caregivers permission not to try to fix you. Remind yourself, and them, that simply surviving is sometimes all we can do. Sometimes all we need is someone to bear witness and keep watch as we heal.

Simplify. Curate your responsibilities, say yes to love, and speak the truth. There will be plenty of time for the work of processing, growth, and whatever your next chapter looks like. Start with simplicity and sit in it as long as it takes. In time, you won't need a life preserver to breathe and will even find your footing on the shore.

Reflection Questions

1. Are you resistant to the idea of assigning positive value to your worth? How do you actually love yourself?
2. Take an honest assessment. What "neighbors" are the most difficult for you to love? What areas of bias are keeping you from showing up for others?
3. What unreasonable or hurtful expectations are you holding on to for yourself or others regarding the grief process? How are these expectations getting in the way of healing?

CHAPTER 9

What Now?

The Discipline of Simplicity

Whether you can see it or not—and often we cannot—everything is being used of God for your good as well as the good of others. This knowledge is by faith. It is not always by sight. But it is nevertheless certain, because it is based on the character of God, who reveals himself to us as both sovereign and benevolent. We are not to be anxious about the unknown future or to fret about it. We are to live in a moment-by-moment dependence upon God.

James Montgomery Boice

There is no question that an internal restructuring is woven into the experience of suffering. When our worlds are turned inside out, many—maybe all—aspects of our lives are targets for realignment. Some of these changes can be unwelcome and uncomfortable. Things that might have once brought us joy might seem dull and empty now. Plans for the future are often rendered obsolete. Finances might require a completely different approach. The world we thought we knew can feel strange and even unwelcoming. Our existing priorities can come into sharp focus, and we might find them inadequate, unimportant, or even harmful. In grief, everything is up in the air, and we don't know how different our environment will look after the landing.

This can feel overwhelming, especially because this aspect of suffering happens with or without our consent. It even happens whether we participate in the restructuring or not, which can result in all kinds of unwanted new norms. Our actual brains change along with the way that we view the world. This can cause a shift in attitudes and behaviors. What if I don't like the new me that is emerging?

When our deep beliefs, or schemas, are contradicted by a sudden change in circumstances, the result can be traumatic. This makes sense. As stated before, because of our loss, we now know something we never wanted to know about the world. We now have to fold that new knowledge into how we understand self and others. On top of that, we are often too exhausted to put forth the energy required to challenge what emerges in the throes of our system realignment. I work with survivors of trauma who often struggle with the reality that they are not actually in control of their safety, or the safety of those they love. We can either submit to this truth with peace, reject the truth in favor of self-blame, or accept it with cynicism and withdrawal. If we neglect to confront it, a more negative outcome is likely. There is a protectiveness in isolation and withdrawal. My brain might slip into that kind of negative reconciliation without my consent if I am not diligent. If you think about it, surely you know someone who has made the determination that not engaging with the world is better than making themselves vulnerable to it.

Nathaniel is a teenage boy who lost a lot in a short amount of time. His family was worried about a change in his expression of affection. They noticed that he hugged and interacted with close family members with the same frequency, but that he was becoming more closed off with others. He was even shutting out family members he wasn't used to seeing on a regular basis. When I asked what his thoughts were about affection, or if he had noticed a change in his behavior, Nathaniel explained it to me in the simplest way. He told me that he would continue to love the people he already loved, but he would never add anyone new to the list. This was the natural response of a precious boy who had lost so much. He knew he had to do something to harness the pain, and his brain presented this solution. No new vulnerabilities. No new chances to be hurt. No new people.

Catherine is an adult client processing complex childhood sexual trauma. Acknowledging the magnitude of the betrayal she endured, and the losses accumulated over time, has her in the throes of grief. The voice she's battling is one that has long been telling her suffering is punishment for her unworthiness. At some point, her brain decided that if she is the bad, broken thing—as opposed to the abuse or the abuser—she does not have to fear the world. If she is at fault, she only has herself to fear. The monster we can see is safer than the monster we can't.

Restructuring

These are just some examples of maladaptive realignments that can define our system changes if we are not paying attention. I have seen plenty of people channel their new way of looking at the world through the lens of unresolvable anger. If we are not paying attention, we too can slip into the norm of feeling constantly victimized by a cruel and uncaring world. In Psalm 13:1-2, even David laments,

> How long, Lord? Will you forget me forever? How long will you hide your face from me? How long must I wrestle with my thoughts and day after day have sorrow in my heart? How long will my enemy triumph over me?

This kind of hopelessness does not have to be our inevitable landing place, though. While certain restructuring takes place in the wake of suffering, it does not have to result in cynicism or dysfunction. If we face these natural tendencies, confront their legitimacy, and consider alternatives, we have a chance to adopt ways of thinking and behaving that are more congruent with who we want to be.

> If we are not paying attention, we too can slip into the norm of feeling constantly victimized by a cruel and uncaring world.

A positive aspect of the process of restructuring includes allowing our thoughts and behaviors to come under the spotlight of scrutiny and challenging the ones that don't align with our values. This is a good thing. In the absence of pushback or challenges from life, we can become apathetic—or at least unwatchful—allowing the world to have more influence on where we put our energy than it should. For example, in my grief, I became acutely aware of how much time I've spent being aggravated over inconsequential things. My tendency to want to resolve every relational kink leaves me vulnerable to seeing only the disconnects and taking on frustrations I have no hope of reconciling. Early on in our grief journey, my husband and I determined that we would no longer "suffer the foolishness" of things that we can't or shouldn't control. This has proven to be no small discipline, and the iron in me consistently needs the iron in him to stay committed to this endeavor. That decision has been, however, a lifesaver. And an eye-opener! There is peace that comes when we carefully curate our responsibilities.

That's just one example of how my priorities have distilled after shedding the barnacles of neglect. In fact, I had to pull way inward in the beginning of my walk through grief. I did not have extra energy to toss to the world at random. I no longer had the luxury of not curating where I spent my emotional coinage. Before grief, I had a lot of me to spare. After the murders, I felt like Bastian in *The Neverending Story*, sitting in front of the empress as she held the last grain of sand from Fantasia in the palm of her hand. Is this all that's left? Yes, but it's enough.

Our suffering might be so consuming that all we have to rebuild from is a grain of sand or a mustard seed. If this is you, it's okay. It is enough. The restructuring, or realignment, we experience as a result of grief looks different depending on our individual experiences. But the idea is the same. What was our normal before gets a dramatic undoing, and we need to regroup.

The first step is being okay with pulling in. It is okay if our worlds get smaller for a bit. It is okay that the focal points of our days look different or are reduced. This is probably not the time to diversify but rather to simplify. This might be all we are capable of anyway. We might as well give ourselves permission to be here. *What is really important to me?* This question might be an explicit consideration or a subconscious wrestling. The profundity of our circumstances in suffering throws us into collision with this idea and begs a response. *What really matters?*

I want to take a moment here to go down a path I think can help with that response. Joshua, a pastor and theologian friend of mine, once introduced me to the spiritual concept of "the second naivete." In *The Symbolism of Evil*, French philosopher Paul Ricoeur describes a transition of faith from "the first naivete," through "critical distance," to "the second naivete."[1] Instead of utilizing his complete philosophical viewpoint, which is worth exploring, I want to borrow his terminology and, in the spirit of this chapter on simplification, attempt to filter these concepts through my own Christian experience of faith.

For our purposes, let's consider the first naivete to be one we might experience as children in church or life. Harken back to what we might have thought of God, of Christians, of the church itself. God has a white beard (it's definitely super soft), Christians are good, and the church is safe. I can picture us all in a Sunday school classroom answering "Jesus!" in unison. And we were right! He is the answer! We blindly accept and trust because this is what we are shown and encouraged to embrace. Believing comes naturally. This simplistic faith, though, might be characterized by a level of ignorance that can't be sustained.

Then life happens. It happens a lot and all over the place, in good ways and bad. We are exposed, especially in this information age, to other ways of thinking and outside opinions about our faith. We are faced with harsh realities about Christians and Christian history. Some of us are wounded deeply by our churches and those in them. We get exposed to the truth that Christian environments don't guarantee safety or care. Does God even have a beard? Our ignorance doesn't stand a chance.

The shattering of the first naivete also occurs in response to the wounds the world inflicts on us. Trauma. Betrayal. Disappointment. Discovery. Being good does not protect us from evil. Being good does not always win us places of security. Treating others the way we want to be treated does not guarantee reciprocal treatment. At some point, we discover that people are playing by different rules and are governed by unseen and unpredictable forces. In this world, it often seems like schemes, plots, and manipulations reign. Unfairness triumphs. Bigotry, oppression, and dehumanization find us and don't even flinch in the face of rationalization. It paints a bleak picture of what we are capable of and what we can succumb to. The wounds of disillusionment matter, and they absolutely affect the way that we view our faith, our relationships, the world around us, and ourselves.

In addition to external disillusionment, most of us face an internal wrestling with self. Ideas like *I am not as good as I thought I was*, or *I will never measure up*, or *I don't deserve to be loved* emerge as a response to the fallenness in and around us. These thoughts are broken but understandable responses to being hurt and hurting others. Facing ourselves in the wake of disappointment is hard. These thoughts can be crippling and can certainly be a catalyst for cynicism and coldness. These negative ideas of self can certainly emerge with a vengeance when we are left vulnerable in the midst of suffering.

And there you have the components of what necessitates a movement from the first naivete to something else. Ricoeur calls this space "critical distance." I think this is the place where we realize that the relationships we thought were indestructible can fall apart. That our heroes are just as human as we are. That despite my best intentions, I am always a hairsbreadth from depravity. We can no longer rely on the innocent but ignorant assumption of fairness. If this language seems harsh, that is because reality is harsh. This world can be cold and savage. There are infinite dangers that we cannot control. We have been hurt. Our trust has been abused. If you really think about it, I'd bet each one of us could make a case for complete cynicism and withdrawal. Give me Walden Pond and leave me alone!

Cynical withdrawal is a valid place to be. There is truth in it. A lot of us land in this place and make ourselves at home. We accept cynicism and negativity in the spirit of protectiveness and realism and make ourselves as comfortable as we can. We figure out how to functionally operate while our brains are bogged down with disappointment and confusion. We convince ourselves that we are better off this way. This kind of withdrawal can be physical, emotional, or cognitive. Even under normal circumstances, no aspect of our lives is without discord or difficulty. Conflict is everywhere, and it complicates the way we interpret our experiences.

Again, we can stay here in the grip of withdrawal, but we don't have to. If we so choose, we can transcend this space of cynicism, or realism, as some might define it. On the other side of disillusionment, the second naivete awaits. Naive perhaps, but not ignorant. Imagine taking everything you have learned about your humanity, about the world, about pain, and funneling it into a new simplicity. Imagine that there is a place beyond bitter acceptance that is characterized by soothing surrender. Again, we are not sticking our heads in the sand and denying reality. We are choosing to go back to the beginning and start asking questions all over again. Who is God? Who am I? Why am I? What does it mean? Do the things I now know about the world change it fundamentally, or reinforce what I started off believing about it? Does it even matter if God has a beard?

Simplicity

In talking to my friend Joshua about this concept, it became clear to me that the idea of transcending disillusionment is applicable across genres. Imagine approaching your relationships through a lens of transcendent simplicity. Maybe our conflicts are not as complicated as I think they are. Maybe every conflict doesn't need to be, or perhaps can't be, resolved. Maybe love really does cover a multitude of sins. Perhaps my happiness or completeness is between God and me and is not my husband's responsibility to provide at all. What if I accept the fallenness in the people I love as a given and not as an affront?

Certainly, our faith journeys transform when we apply a theology that is distilled and simplified. It is not a stretch to imagine that this might be exactly what Christ meant by a childlike faith (Matthew 18:3-4) or what Paul was referring to in Romans 12:2 when he urged us not to "conform to the pattern of this world, but be transformed by the renewing of your mind." Paul goes on to say that it is in this place of rebirth that God's perfect will is discernible.

Let's consider Paul's appeal. No doubt part of the pattern of this world is defeat, bitterness, accusation, pessimism, narcissism, hedonism, and isolation (refer, once again, to Psalm 13). We can assume, then, that renewal involves arriving at a place characterized by the fruit of the Spirit: love, joy, peace, patience, kindness, goodness, faithfulness, gentleness, and self-control (Galatians 5:22-23). *Simplicity.* How, then, do we embark on a renewal, or attempt to transcend critical distance?

First, I think it is important to be able to conceptualize that this is a possibility. We are capable of climbing out of the fog of betrayal. Further, spiritual transcendence is a desirable destination. There is something on the other side of disappointment and defeat. A newness is possible. We can consider our movement in life and in grief to be pointed toward this newness. Thank goodness disillusionment is not the end of the story. Thank God there is more to see and learn!

There is something on the other side of disappointment and defeat. A newness is possible.

Swiss psychologist Carl Jung argued in favor of this kind of work when he described "the second half of life" in his 1939 essay, "The Stages of Life." Jung and Jungian clinicians suggest that this stage involves finding meaning in life and death, prioritizing values and relationships, abandoning external validation for inner fulfillment, and embracing uncertainty. This is deeply spiritual work. According to this quotation accredited to Jung, "the first half of life is devoted to forming a healthy ego, the second half is going inward and letting go of it."[2]

For Christians, this might be akin to the idea of dying to self (Luke 9:23; Romans 6:1-11). The less of my ego there is, the more space there is for Christlikeness. Denying self here means we have some letting go to do. We are entitled, in a sense, to our cynicism. Most of the time, it's hard won. We can definitely wrap the bitter truths we've learned around us like armor and dare the world to get close. We can exist in a state of aloofness, only entering shallow relationship and interaction. The challenge is to let go of some of that safe, critical distance and commit to vulnerability. Embracing uncertainty, dying to self, and committing to vulnerability are scary endeavors when we've been wounded. Ricoeur, Jung, Paul, and Christ seem to agree, however, that the path to peace is simplification.

My mother-in-law, Karin, lost her husband of 49 years to cardiogenic shock in April 2022. She recently asked my thoughts about what it means to lean on God when moving through grief. This can be a complicated question, especially when grief has us confronting our own thoughts about life, death, faith, and God's character. As previously discussed, our priorities and worldviews are up for realignment as a response to suffering. In light of the notion of simplification, we might first refer back to chapter 5 of this book and remind ourselves that there are things we need to relinquish in order to get out of God's way.

Second, let's take a minute to consider the word *lean*. Probably the most well-known scripture regarding leaning is from Proverbs 3:5: "Trust in the Lord with all your heart and lean not on your own understanding." In this passage, *lean on* is synonymous with *rely on* or *trust in*. Another interpretation is to *rest on*. This last meaning, in particular, is intriguing. We are thinking, reasoning creatures and can't help but construct, deconstruct, and reconstruct what's on our minds. I even try to make sense of things and solve problems in my sleep.

(You can ask my husband, who sometimes has to listen to my sleep-talking dissertations.) We are all Sherlocks, sleuthing out the facts, piecing them together, hoping to achieve a triumphant denouement. And then we can rest on it—rest on what we've figured out. There is relief in resolution.

If you refer to the chapter on forgiveness, you might remember the idea that we are woefully unequipped to be able to fully assess a situation. In other words, I might be able to achieve a thorough understanding of my own experience, but the whole picture is out of my reach. This is by divine design. We are not given access to all the pieces. God's revelation falls under the domain of His sovereignty. If we knew everything we needed to for peace and security, faith would lose its function. The four-dimensional context of a situation is impossible for us to comprehend in totality. While my brain wants to lean on my own understanding, the discipline of remembering God's sovereignty requires me to relinquish that surety to Him. That does not leave us with nothing substantial to lean on. That does not mean we don't get to rest. In fact, this kind of submission is where true rest really is.

The imperative to find true rest in submission might be more acute in trauma, grief, and suffering than in any other human experience. It's where the rubber meets the road. When things are going well, it is easy for me to lean on God. My environment isn't at odds with His goodness! It is when the world around me serves up something that seems contradictory to who God is or who I want Him to be that I run into trouble. We might pray, *Lord, leaning on You means relinquishing control. Giving up control in this context means taking everything I know about the world, self, and others, and submitting to the understanding that these things are simply tools to help me see You better. They are tools to help me submit and respond to Your will.*

Yikes. This, in the midst of my panic? This, while my subconscious need for control is in overdrive? When I have lost something irreplaceable, am I really supposed to give up more? The short but complete answer is yes. In Philippians 1:21, Paul says that "to live is Christ and to die is gain." Further in the chapter Paul is clear that he believes that the glory that awaits us in heaven is far greater than what we experience on earth (verse 23). So gain in that sense is a very literal gain of eternity in heaven with God. But the language also refers to the idea of dying to self and the earthly gain that comes with submission to Christ.

This is the ultimate beauty of living in the second naivete. The gain here is peace. Hope. Connection. Vision. Simplicity. In the seeming impossibility of ever finding happiness again, we gain access to joy. Through my own lived experience of laying down my notions of control at the feet of Christ, I have been able to find fulfillment in the simplicity of His goodness.

God is either everything He says that He is, or He doesn't exist. There is no in-between. Since I know He exists, my job is a lot easier than I thought prior to intense grief. There is so much rest in the knowledge that it is all handled and handled perfectly. God does not need me to sleuth it all out or force pieces to fit. He certainly does not need me to take over. My job is simple. Trust in Him and actively live in His creation. Be a good steward of what He has put in front of me. That means tending to others with love, taking care of my mind and body, accepting and honing the gifts He has bestowed on me, and having these things prepped and ready to be used for His purpose.

> We can find fulfillment in the simplicity of His goodness.

Curiosity

Finally, this great act of reorganization and simplification addresses an aspect of trauma that we don't always discuss: the wounding of curiosity. *Anhedonia* is a term that refers to the absence of joy in things that typically evoke it. These things are often collateral damage when we have been wounded and are dealing with the long-term effects. Our desire to engage with the world can be so greatly compromised that we can wonder if there is even a point. You are not alone. While there are several avenues, including counseling, you can use to address this particular aspect of PTSD, depression, or complicated grief, simplification helps. Pull in, reassess, and begin the discipline of letting go. Remember the time you spent in your "first naivete" when the world was mysterious and magical? Do you remember when everything felt possible? When was the last time you were awestruck, undone by the wonder of the world? When was the last time you engaged with creation with delight? This is what awaits us in the renewal of the mind. The childlike faith that emerges with simplicity allows us to experience curiosity and wonder in a whole new way.

I want to take a minute here to acknowledge the impact community can have on this process. Refer to chapter 4 of this book on the necessity of investing in your relationships. Those in our inner circles can vitally assist us as we try to simplify and reorganize our priorities. They can help by giving us permission to change, allowing us to borrow some of their energy, bolstering our resolve, and simply making themselves available as listeners. We are created to be in relationship. The hard work of evaluating self and the world is more easily done in the context of trusted partnership. It is an adventure made richer by witness.

In the space where everything you've learned and all you know exists, simplicity can be the spark of a new way of being. The good news of the gospel is profound, uncomplicated, and all we need. In addition to being relieved of so many earthly burdens of control we have adopted, we find something else. Something shiny and new. Our eyes, once so aged and weary, can see the world with a childlike glimmer again. There God waits for us with renewed gifts of wonder, awe, and mutual delight. This is the state of mind where Nathaniel, closed off to new people because of his great loss, learns to love again. Catherine, grieving childhood sexual trauma and feelings of unworthiness, accepts her intrinsic value. And we, while strolling with Jesus, find new gardens.

Reflection Questions

1. What new thoughts and ideas about the world have emerged in your journey with grief? Are they helpful? Honest? Are they congruent with your values?
2. Have you experienced a "disillusionment" with regard to your value structures (like the church)? What was that like for you, and how have you responded to the change in perspective?
3. How has your grief affected the ways you enjoy life or the world?

Conclusion

Looking to the Future

To develop a relationship with God through Jesus Christ is your number one job and challenge. He will make Himself and your path powerfully and clearly known to you. In Him, you will find every answer to every question—both big and small. And through Him you will never have to try to make an impact—you just will... by being you.

Barbara Lesslie, in a letter to her granddaughter, Caris, on her sixteenth birthday, March 28, 2021

I am a mountain woman at heart. The mountains sing to my soul, make me feel alive, and hum with a gravity that reminds me that there is more out there. Standing at the water's edge, bearing witness to the ocean's rhythmic conversation, has the same effect on me. Nature points to something bigger and better, something mysterious and good. It evokes our deep connection to creation itself. Nature catches us up in the promise of something more. Knowing what we know of the character of God, we can be certain that this is by design. God calls to us through nature, using it to point to the truth of who He is and the certainty of what lies ahead.

My soul longs to connect to that promise of scriptural eternity. The natural world whispers that promise to me, as does the nature of love. Connection through relationships gives us glimpses of what is to come, when we get to be in perfect communion with God's creation. Human nature, no doubt, complicates the message. The fallenness of the world breeds damage. Discomfort sets us off at every turn. We crave gratification in all forms. We hurt one another to get that gratification, and the resulting wounds require tending.

Navigating pain and suffering takes on different shapes and expressions. The journey is sometimes abrupt, loud, or declarative. It can be messy, emotional, or labor intensive. It can look like major lifestyle changes, public advocacy, or proclamations of insight. It can look like writing a book. Movement through grief can also be so subtle it is undetectable from the outside. These might be the grittiest moments of suffering. Moments where all I can muster is a gesture that maybe no one else can even see. These are okay, and every bit as important as the grand demonstrations of intent. At times, all movement needs to be is facing a direction. Pointing to something. Whether forced or initiated willingly, these small, yet worthwhile intentions are necessary.

There is depth in this simplicity. Pointing to something indicates that there is somewhere to go, someone to be, something beyond where I am. Facing a direction means that I am choosing to acknowledge that there is, in fact, a future to look toward. Hope. This is no small thing. As we have explored in previous pages, grief can leave us uncertain that a future under our specific circumstances is survivable or even imaginable. We might not want a future where our story is reality. We might be so exhausted that consideration of the future is too much to muster.

Much like the voices of the misty blue mountains of the Appalachians or the shores of South Carolina beaches, God is calling. Remember the idea of the vessel from chapter 3. Consider the concept of worldview. No doubt grief leaves uncertainty where clarity of direction used to be. What can I pour that uncertainty into that will yield peace? Scripture reminds us that God's daily provision is enough. Exodus 16 tells the story of the Israelites sojourning through the wilderness. They became disgruntled and impatient, even longing for the supposed security of slavery! God's response was manna. He provided for His people a daily bread that would sustain them. He even told them specifically not to store up food for tomorrow, reminding them of his consistent presence and complete provision.

My dad, whose deep faith is a beacon to me, had a hilarious amount of Irish Spring soap stored up in his home. When my siblings and I were sorting through my parents' things after their deaths, we just kept finding more. In an effort to be good stewards, I think we all silently committed to using every last bar of soap. We split the collection four ways and, to the best of my knowledge, are still working through them today, almost four years later. While dad's commitment to making sure he was never caught without a bar of Irish Spring is endearing, it speaks to a human problem I find so relatable. If God is blessing me with manna today, I want to grab as much as possible just in case. I want to hoard the good things, just in case things change, just in case I am blindsided, just in case tomorrow is the day I lose everything. We are human. We know that life is not fair and suffering is guaranteed. *Lord, let me insulate myself as much as possible from the pang of want.*

Numbers 11 recounts the story from Exodus 16. The chosen people God had delivered from the injustice of slavery found themselves faced with a different kind of uncertainty and discomfort. They had enough bread but longed for meat. The English Standard Version of Numbers 11:4 says "Now the rabble that was among them had a strong craving." The masses grumbled. I deeply appreciate this translation. It is so simply stated but so universally experienced internally and from a modern-day cultural standpoint. Yep! I am the rabble! Now, I am not proud of being the rabble, but I have to admit that I have strong cravings and will cause a ruckus to get my needs met. My response to longing is to seek gratification from the world around me. What keeps me from hurting others in that process?

This is where we can take our vessel, our worldview, and turn to our people for reinforcement. The relationships I have either invested in or are starting to invest in are my source for compassion, correction, and love. If I am choosing wisely, these are my people who will tell me I don't need that extra bulk order of Irish Spring. They are the ones who help me stay aligned with my values. These are the relationships that will remind me that the beauty of faith is that it's enough to hold even the most rabble-inspired aspects of my uncertainty. When I need energy to be steadfast, these are the arms I will lean into.

The connection here is important. My worldview will determine how I build my community and who I fill it with. My values dictate the kind of voices I surround myself with. Those voices, then, can build me up, reminding me who I have chosen to be, or they can allow me to abandon myself. They can either provide comfort and wisdom or make me comfortable in my foolishness. I want my people to see the rabble in me and remind me of God's provision. I need them to bring me a mirror and support my spine while I look into it. I want my people to feel empowered to be able to lift my arm to point toward the future. I can invoke the Spirit of discernment in this endeavor. I have the power to choose who those people are and how available I am to their intervention.

My propensity toward rebellion, self-reliance, and stubborn foolishness can lead me down a cyclical pattern of disappointment, much like the people of Israel. I need to seriously consider how I approach the future I want for myself, post-grief. First, I need to realize that I cannot control it. If I could, I wouldn't be in a place of pain and suffering to begin with. I am my father's daughter. I want to hoard the soaps. I want to protect myself and my loved ones from famine and disaster. Will I trade that false sense of security for the beauty of promise, though? God told Israel that reliance on Him means laying down their earthly sense of protection.

This is especially hard when we are combating understandable and often unavoidable trauma responses. My fears flare up when the world delivers something my flesh interprets as dangerous. I can't help it. What I can do, however, is work to make sure I have resources in place that can settle my nerves, speak truth and wisdom, and calm the internal rabble. Looking to the future with faith means getting out of God's way and trusting in His provision. Simply speaking, it takes installing constant reminders that God's purposes and plans are better than mine. If what I know about God is true, and it is, I can live with Him in control.

Part of that discipline is tackling the potential stumbling blocks I have most likely nurtured in my life. What holds me back? What trips me up? These are the questions we need to be asking ourselves as we point to tomorrow. Solution-focused therapy is a modality that is pretty much as simple as it sounds. Keep doing what works and stop doing what doesn't. This is a good philosophy to apply while honing this discipline. If we are being honest with ourselves, we typically know what is helpful in our lives and what isn't. Up for grabs here, for example, are lifestyle norms such as substance use, exposure to toxic media, and allowing negative influence from certain cultural or social groups. Am I exercising? Eating? Sleeping? Face it, there are some obvious things we can tweak here to better support the steadfastness required to move forward in faith.

We also have to look at emotional and relational norms. How am I handling conflict? What has become my typical response to being told no or being confronted about my behavior? Have my tendencies to react dysfunctionally gone unchecked? When we are dealing with grief, this can be very hard to discern. In a sense, our suffering entitles us to our dysfunction, and our people will often support it if it means not causing us any more discomfort. It is hard to do, but we might need to give our support systems permission to do an independent assessment of our grief responses, especially after some time has passed. If we don't empower them to intervene when necessary, they may fall back on the often invalid assumption that we know what we need and what we are doing. We can easily become people who require enabling from others, and we can become prolific enablers of others. Again, this is why being an active participant in relationship investment is so important. We need our people to sharpen us with love.

One resource we have as a consideration for self-assessment is counseling. The work described above can be very difficult. Looking in a mirror closely is rarely fun and typically requires an active response. If you do not have people who feel comfortable holding you accountable for your dysfunction, find a therapist who will. When it comes to counseling, a good fit goes a long way. My recommendation is to find a counselor who is trauma-informed, adept at holding hard things, and not afraid to offer challenges. Interview them and don't hesitate to move on to the next one if you don't feel comfortable or like you are in competent hands. Remind yourself that you are not in therapy to stay the same. Counseling is not designed to be an echo chamber that enables you to dig deeper into dysfunction. You are not shopping for someone to cosign your maladjustments. An important therapeutic goal here is to let go of what is holding you back and release what you are using to prop up a false sense of control.

This is the time to consider how unforgiveness and regret are serving as stumbling blocks. If my goal is clearing the way so I can point to a new thing, I likely need to confront the old things that are hanging on. As explored in chapter 6, forgiveness is not a one-off. It requires humility and discipline. It's a long road. God spends a fair amount of time on forgiveness because it is good for me. Ultimately, it helps me see God more clearly. It reminds me of my own fallibility. It is good practice for letting God guide my steps. If I am going to have the energy to face tomorrow, it will be helpful to free up the energy I am spending on unforgiveness and clear the debris.

The same holds true for what we typically think of as regret. If regret looks like self-flagellating rumination, I need to unpack and wrestle with it. If regret is tied to shame and forcing me into hiding, I need to shine a light on it. Regret functions differently when tied to guilt. It becomes functional. Regret in a functional sense serves as a reminder of lessons learned. It straightens my path and honors the mistakes I have made as tools for refinement. As much as shame is my adversary, guilt and regret can be my friends. Self-forgiveness and growth are inherent in this particular wrestling. Freedom emerges if we embark on it with honesty and accountability.

This kind of freedom has consequences: compassion. The space I have just created in my heart and head gives way to a new kind of engagement with the world: connection. If I can settle the turmoil of unforgiveness and shame, I can feel the nudgings of the Holy Spirit more acutely. As we navigate suffering and attempt to point toward the future, we have important choices to make. Do I want the wages of my pain to be hidden in isolation and withdrawal? I certainly get to choose that. Do I want to repay the world's unfairness with coldness? Who would blame me? *Lord, haven't I done enough?* To suffer is to toil. To suffer greatly is to bind parts of your soul to the effort of surviving. The temptation to consider our sacrifice complete is considerable. Hasn't the torrent of my tears secured my heavenly crown? Can't I just sit the rest of the hard stuff out? Thank God there is another option.

The other option is engagement. Entering in. Loving your neighbor. Only, now I can do it so much differently. I watch the way people look at my brother or approach my sister post-tragedy. I see people lean in to hear more, see more. They are looking for hope. They want to know that it is all real: the possibility of survival, the ability to carry on, the capacity for laughter, goodness, and God Himself. People want to tuck away the knowledge that when they are faced with suffering, they too will survive. While I cannot claim to embody all the positives of post-traumatic growth, I can certainly bear deep witness to the truth that is the gospel. This is what we get to do when we enter back into the world after loss. Bearing witness is a holy and sacred endeavor.

We might be entering back in like a baby giraffe, all wobbly legs and awkwardness. We might feel like a gray cloud, filled to the brim and ready to rain down at any moment. We might have no idea what to say or do. I don't think that matters. What matters is that we show up, offering our love through the struggles and gifts that have been bestowed upon us. What matters is that our experiences bear witness to the truth for which others are thirsting. We enter in, we engage, we connect.

When I was barely standing on my own feet, sadness and bewilderment swirling around me like chaos, I had to go to the local post office to address issues we were having with mail thieves. In situations where deaths are published in the news, identity thieves jump into action. It was a devastating realization for my family. Trauma on top of trauma. To say that I was raw and frazzled at the post office on that day is an understatement. At the counter next to me was a beautiful, soft-spoken woman. I remember paying attention to her because, like me, she seemed to be filled to the brim with raw emotion. She had several packages to mail and explained to the postal worker that she was in the midst of grief, was emerging from isolation, and that she had begun to engage with her art again. These packages represented her first shipment of art into the world since experiencing loss. I don't think she told the postal worker this because they asked (although they are very kind at my post office!). I think she needed to say it out loud to declare it.

She and I finished up at the same time, and as we walked toward the exit, I put my hand on her arm and shared with her that I too was grieving. I told her that I was so grateful that she was reclaiming her art and was sharing with the world. I needed to see a light, a glimpse of the path further down. This stranger grabbed me gently by the shoulders and began to pray over me. As she prayed, her arms wrapped around me, and we cried together. When she finished, she looked at me in the eye and told me she loved me. It was the most natural thing in the world for me to reciprocate. I meant it then and I mean it now. I do love her. I don't know her name, or what happened to her after that encounter, but I love her. I think about her quite a bit.

This is love. She went out into the world, into the post office, not knowing who was going to need her, and she answered the call of the Holy Spirit to connect. It's something my mom would have done. She was so good at feeling the prick of need from others and leaning into it. People felt loved by my mother, and they were right to feel that way. I want to be that way too, like my friend at the post office, like my mom. My way might not always be post-office-hug-sessions (although I am here for it), but it is up to me to connect with my particular gifts and use them to reach others.

I want to be the hands and feet of Christ. It is this simple. If I trace the lines of all I know down to the cornerstone, it is that God is who He says He is. There is a lot of life, pain, disappointment, and confusion that certainly piles up on top of that foundation. It is not a simple thing to deal with all that. The ultimate answer, however, is simple: Jesus. He is the answer to my pain, my confusion, and my weakness. He is the answer to the great question, "*Why*?" Grief is indeed threatening. It is the weapon used against Job. It is the quicksand that undermined

David's dynasty. It is the meteor that shattered the world as I knew it. Grief is also a conduit for hope. It is a pathway that leads to meaning and depth. In it is the thinning of the veil.

In the dance between darkness and light in suffering, raw, simple truth can be found. In this place, we are invited to wrestle with ourselves with the hope of emerging with purpose and meaning. It is a place where God is waiting. Your pain is not without heavenly witness, not without divine purpose. What you do with that understanding is the next chapter.

> Your pain is not without heavenly witness, not without divine purpose.

Reflection Questions

1. How could you be getting in the way of God's daily provision by "hording blessings?" How are you using control to guarantee your safety instead of relying on God?
2. What lifestyle choices of yours need an adjustment so you can be better equipped to navigate hardship?
3. Which gifts of yours can you use to better minister to your neighbors? How can you encourage and empower yourself to be more likely to make those choices?

Voices from the Couch on Suffering and Grief

Wisdom from My Family, Friends, Clients, and Colleagues

The following offerings are compiled in an effort to give voice to other people's experiences of grief and suffering. Experiences are best relayed through the very particular and unique language of the people going through them. I hope that some of these voices resonate with you, and you find encouragement that you are not alone.

“Music and grief—sometimes solace and sometimes heartbreak. Bagpipes playing 'Amazing Grace' can be both. But the hymn 'Just as I Am' brings instant tears. Tears can be a welcome release.”

—*Patricia, septuagenarian*

“When you lose someone so precious to you, you come to realize that they were not yours to begin with. They were a gift from God. The most difficult thing is to praise God in the taking as I did the giving.”

—*Ann*

“In thinking about grief I am reminded of the four most important statements that can be spoken as one approaches the end of life. Please forgive me, I forgive you, Thank you, and I love you.”

—*Robert*

“Grief comes in waves—some gentle, some crashing—but if you ride them instead of resisting, the journey shapes you into someone stronger, wiser, and more deeply human.”

—*Kristin*

“I’ve been thinking lately about grief ignored. Grief that we deny, even to ourselves, can still change who we are. Ignoring grief won’t make it go away. To be human, to love others will bring grief and suffering to each and every one of us. And in that grief we have an opportunity to build connection with one another if we choose love, trust, and understanding.”

—*A friend*

“The world doesn’t stop spinning because you are sad. You’ll find that over time, grief never heals; it becomes a part of who you are, and that makes it easier.”

—*Caris*

“Grief teaches you to love by faith and not by sight. I have learned to appreciate grief from a spiritual context and to dwell in the suffering even though my heart is breaking.”

—*Grandmother*

“My grief is God-given. My grief is uniquely my own, which makes it incomparable to your grief. Our grief makes us part of a club we did not choose to be in; it is a club where everyone is equal. My grief allows me to see joy through sorrow, to look for the light in the darkness, and give grace to myself and others who are grieving. My grief is not linear; it is cyclical. There are days it is okay to not be okay. My grief knows that I don’t always have the words for your grief, but I see it nonetheless.”

—*Lori*

“*Pray*...even when you’re not sure if God is there. *Exercise*...even when you don’t feel like it. God gives us those endorphins for a reason.

Laugh...even when you think you never will again. Watch your favorite comedian on YouTube!"

—Lucille

"One year ago, I wrote a post about grief. My grief for the church that I thought was about to close. It felt so heavy. A few days later I experienced and witnessed a grief that was...well that was...that was indescribable. A grief that made a post about a church closing seem so ridiculous that I deleted it. And in that grief I became a witness to the most beautiful expression of God's grace and love. A love that can only come from God. It feels horrible to be grateful for the thing that I always wish I could take away. But grief and gratitude can be so closely knit together. 'I see you, grief. I am grateful for the unbelievable love I had that is connected to you.' And in a few days our church will have its last service. On Easter. We will declare...in the worst darkness, there is always hope."

—Grief one year later

"One step, minute, hour, day, week, month, year, decade at a time."

—Dave

"The first time I ever recognized sadness in a stranger's eyes was after grief had left its mark on mine. If heartache hadn't stopped my world, I might never have been still enough to notice. I think that's what grief does, slows us down just enough to see others more clearly."

—Bella, at 19, the year I learned true compassion

"We are all like the moon. We all have to go through some phases of darkness so we can shine our brightest."

—Connor

"Obviously, the emotional toll of (my husband's) death was (and still is) overwhelming, but I certainly wasn't prepared for the physical toll—it's as if my mind decided that grieving was not enough; I had to suffer physically also. One of his ICU nurses (about nine months after his death) asked me if I got out of bed every morning. I said, 'Of course,' and she said that was enough—I was doing fine."

—*Karin*

"Grief is the space between what is and what we hoped it would be."

—*Patricia*

"No matter how hard you may fight it, grief fundamentally changes who you are. It leaves behind a new version of yourself that you will have to relearn and make peace with."

—*Suze*

"When one is young, grief is so new, unknown, and even unexpected. It is like a thief in the night that only and suddenly brings loss and sorrow. But the more one lives and experiences grief, the more the person of faith comes to see that grief is more like an acquaintance that comes around every so often—an acquaintance not to be hated or despised—but one that points toward and reminds us of God's truth of Life in this world of death."

—*Barry*

"Grief is a thief. A grifter. It wants to steal from you. But like an online con artist, it needs your participation to do it. If you cooperate with the scheme, it can rob you of your time, your trust, your joy. A little at a time, day by day. Initially, grief's true goals are concealed, and you don't question its intent. But in time, grief reveals its true nature—like the toxic frenemy who secretly undermines you. And once that happens, you can take control and minimize grief's ability to take your happiness."

—*Otto Graff*

"The depth of my grief fluctuates from day to day. Sometimes it is all-encompassing and sometimes imperceptible. Grief at times feels like walking in quicksand with weights tied around my ankles."

—*Grieving the loss of grandchildren*

“God does not always give us answers but He gives us Himself and that is enough.”

—*Jeanie*

“If you continuously create a story of how your life should have been without your loss, you choose to continue to suffer. Ultimately it is not the grief that stops you from starting life over, but the fear of losing that life all over again.”

—*Ann*

“Be kind to yourself. Some days you won't be able to be with people who are happy.”

—*Grief of a mom*

“Grief is like the ocean. It comes in waves that can knock you down when you least expect it. When you face the waves you are able to prepare for the impact, and by looking beyond the waves to the horizon, you can see calmer water and that provides hope.”

—*Sarah*

“Grief, shock, disbelief, anger, spiritual unrest, and questions of faith. Realization that physical, temporary life on earth is but a vapor.”

—*Ken*

“With love comes grief and pain, but a life without love is unfulfilled.”

—*Denton*

“Two key practices helped me during the grieving process of my father—acceptance and gratitude. First, I had to accept that God always knows more than I do and that his death (and anything for that matter) occurred exactly how and when it was supposed to. Furthermore, if I am to enjoy life and at all its beautiful moments, I must be willing to accept tragedy and loss. This understanding helped me move toward a state of gratitude—how fortunate I was to have experienced such a significant relationship with someone that the loss of them in my life could have such a profound effect."

—Becka

"Grief is not a place to stay, but something you work through."

—Anonymous

"My unsolicited advice about grief? Figure out how your personal grief fits with the grief of those most important to you. No one grieves the same; therefore, no one will follow the same path. I was blind to how my personal grief affected my inner circle. I selfishly and erroneously neglected the fact that those around me did not feel the same as I did. When you endure a hardship together, figure out your different roles in navigating it together."

—Lori

"The grieving are often plagued by unspoken pressure to disguise their pain and make it more digestible for others. The grieving should not bear the responsibility of making others feel comfortable around their pain. The only job of the grieving is to grieve."

—Alissa

"There are no short cuts through grief. There are no turnarounds either. The only way to go is through, one day at a time, and sometimes, one hour at a time. No one else gets to rush you. No one else gets to tell you how to do it or when to be done grieving. It just takes time."

—Bev

"What the heart loves best, it remembers most."

—A therapist

"It's true, your mind is not playing a trick on you, life is continuing all around you. You have to find the way to do the same."

—Grieving the loss of my dream

“Grief sucks but it means that you’ve loved.”

—Christian

“So often it seems the deep ache of grief is only detected by others who have experienced it in some way.”

—Grieving infertility

“It takes courage to grieve. Give yourself permission to do so in all the ways your heart needs.”

—Someone trying to be brave

“There is always someone who needs a helping hand. That hand might be the one that pulls you out of the pit.”

—Anonymous

“I think what I am learning about this process is that there are a lot of pitfalls and stumbling blocks to grief. It is not a linear process. It is a whirlwind of twists and turns with stumbling blocks and pitfalls at every turn. Every time you fall into one of the pits or trip over one of the blocks, you have to give yourself a huge dollop of grace and try again.”

—Grieving the loss of security

“The road of grief is often paved with isolation.”

—Anonymous

“Grief is yearning and aching and wanting to go home.”

—Julie

“Live in the present. You will see them again in due time.”

—Someone picking up the pieces after losing another parent

“Write what their life meant to you and what their loss has meant for your life.”

—Amanda, missing my friend

“Grief to me is like screaming in an empty room. It is an endless wave of echoes that knocks you to your knees, but no one truly hears it until they also enter.”

—*Melissa, grieving the loss of a son*

“Grief is too simplistic a word to encompass this season of life.”

—*Julee*

“Grief is like the weather...always changing and often unpredictable.”

—*Someone waiting on the sun to come out*

“Grief is lonely, but you don’t have to do it alone.”

—*Been there*

“Grief is fearfully facing the pain of a shattered childhood. It’s the slow and painful realization that you were not loved and protected by the people you trusted most. But thanks be to Him who heals the broken hearted, binding up our wounds and healing our heart. Only our Savior can mend the broken places in us.”

—*Mary Catherine*

“You cannot take away someone else’s grief; you can only help them carry it.”

—*Anonymous*

“For me, tears of grief help to wash away some of the pain. It cleans out a space to hold the love that we’ve been lucky enough to experience.”

—*Deirdre*

“Grief is hard. But as my Papa would say, sometimes all you gotta do is a woot and a toot.”

—*Bits*

“For me, grief has been like a deep cut that over time, heals and scabs...but is easily reopened, often unexpectedly. Slowly, slowly, it heals again...but the scar is always there below the surface and I never know what will reopen the wound. It's also a sacred space...where only God can truly heal. It's an unspeakable privilege to walk beside someone in the valley."

—*Connie*

"The greater the love the greater the grief."

—*Wayne*

"Time doesn't heal all wounds, but it makes it easier to live with them."

—*Grieving the loss of a friend*

"Grief comes from the heart, not the head."

—*Dipwad*

"Here's what I keep in mind: 'There are many rooms in my Father's house. I would not tell you this if it were not true. I am going there to prepare a place for you. After I go and prepare a place for you, I will come back. Then I will take you to be with me so that you may be where I am. You know the way to the place where I am going...I will ask the Father, and he will give you another Helper. He will give you this Helper to be with you forever. The Helper is the Spirit of truth' (John 14:2-5, 16-17 ICB). The Holy Spirit is real and the future is bright."

—*Robbie*

"Grief is God letting you experience the full weight of loving someone. Anytime you feel it, you know it is a gift. Without love, there would be no grief. God often speaks the most clearly through loss, since He will never leave."

—*Austin*

"They are more than the day they died."

—*Bits*

“Seeing the ‘gifts’ of their memories are stabs of loss. But also an invitation to receive the gift again.”

—*Walt*

“So many people reach out for you during grief. It’s appreciated, but a part of you feels your life is irreparably different from the day before and theirs isn’t. It makes you feel like you don’t match. Wear the clothes you’ve been given until you find a way to fit into them. God is near to the broken hearted and the people who don’t match.”

—*Wendy*

“Don’t try to avoid it. Don’t go through it alone. Grief comes in waves and can hit you out of nowhere. Face it, feel it, let it out, stand back up, and keep moving forward.”

—*Tammy, someone still recovering from the trauma of an event that makes you question your entire life*

Reflection Question

What are you learning through your own experience with grief and suffering? What thoughts can you add to this list of voices?

My Reflections

NOTES

Chapter 1: On Suffering and Grief

1. HC Saavedra Pérez, et al., "Cognition, Structural Brain Changes and Complicated Grief. A Population-Based Study," *Psychological Medicine* 45, no. 7 (2015):1389-99, doi: 10.1017/S0033291714002499. Harald Gündel, et al., "Functional Neuroanatomy of Grief: An FMRI Study," *American Journal of Psychiatry* 160, no. 11 (2003):1946-53, doi: 10.1176/appi.ajp.160.11.1946. SE Kakarala, et al., "The Neurobiological Reward System in Prolonged Grief Disorder (PGD): A Systematic Review," *Psychiatry Research: Neuroimaging* 303 (2020), doi: 10.1016/j.pscychresns.2020.111135.
2. C.S. Lewis, *Mere Christianity* (Macmillan, 1977), 52.

Chapter 2: Reeling from Impact

1. World Health Organization, "Acute Stress Reaction," *International Classification of Diseases,* 11th Revision (ICD-11), 2018). American Psychiatric Association, "Acute Stress Disorder," *Diagnostic and Statistical Manual of Mental Disorders (DSM-5)*, 2013.
2. American Psychiatric Association, *Diagnostic and Statistical Manual of Mental Disorders*, 5th ed. (American Psychiatric Publishing, 2013).
3. Bessel van der Kolk, *The Body Keeps the Score: Mind, Brain and Body in the Transformation of Trauma* (Penguin, 2015).

Chapter 3: The Vessel

1. Although this is my definition, I've drawn from the ideas from James Sire's book *Naming the Elephant: Worldview as a Concept* (InterVarsity Press, 2004).
2. James Sire, *The Universe Next Door: A Basic Worldview Catalog* (InterVarsity, 2020).
3. The Cure, "A Letter to Elise," on Wish, Fiction Records, 1992.
4. Erik H. Erikson, *Childhood and Society* (W. W. Norton & Company, 1950).
5. Erikson, *Childhood and Society.*

6. Sire, *The Universe Next Door*, 8.

Chapter 4: Help

1. Henri Nouwen, *Community* (Orbis Books, 2021), 59.
2. Matthew 4:1-11 and Luke 4:1-13.
3. Luke 5:16, Mark 1:35, and Matthew 14:23.
4. God sets boundaries in the Garden of Eden even before the fall with the mandate to abstain from the tree of knowledge (Genesis 2:17). Subsequent boundaries are found in the form of the Ten Commandments (Exodus 20 and Deuteronomy 5) as well as in the Old Testament Law. Jesus sets boundaries for personal space (Luke 5:16; Mark 6:31) and regarding specific behaviors and relational attributes through His parables.

Chapter 5: Making Room for God

1. Alison Gee, "Who First Said, 'The Pen Is Mightier Than the Sword?'" BBC News, January 9, 2015, https://www.bbc.com/news/magazine-30729480#:~:text=The%20English%20words%20%22The%20pen,up%20arms%20against%20his%20enemies.

Chapter 6: Tackling the Unthinkable

1. Quote Research, "Quote Origin: Forgiveness Is the Fragrance the Violet Sheds on the Heel That Has Crushed It," September 30, 2013, https://quoteinvestigator.com/2013/09/30/violet-forgive/.
2. Alexander Pope, *An Essay on Criticism* (Lewis, 1711).
3. Charles Haddon Spurgeon, "Grace for Grace," *Sermons from Metropolitan Tabernacle Pulpit*, vol. 35 (Banner of Truth Trust, 1970).

Chapter 9: What Now?

1. Paul Ricoeur, *The Symbolism of Evil* (Beacon, 1967).
2. Carl Jung, "The Stages of Life," in *The Collected Works of C.G. Jung*, vol. 8 (Princeton University Press, 1970), 387-403.

Acknowledgments

Writing this book over the past several years has been an intimate part of my grief journey, providing space for sadness, reflection, hope, and love. It could not have happened without the support of my husband, Dave, who has stayed by my side, in and out of the trench, through it all. My children, my siblings, and their spouses have provided me with grace, laughter, challenge, and infinite reasons to push through the hard for the sake of connection and love. My friends have encouraged and celebrated my efforts to get this done. Harvest House took a chance, invited me into the fold, and expertly held my hand through each step. Finally, I want to thank my mother-in-law, Karin, who, as of this writing, is the only soul who has read *In the Valley of the Shadow* in its entirety. Karin, thank you for entering in, engaging with this difficult topic, guiding me, and combing through the manuscript for grammatical errors and clarity. Your invaluable partnership has made me a better writer. Thank you.